Ordinary Geniuses

Ordinary Geniuses

Max Delbrück, **George** Gamow,
and the Origins of Genomics and Big Bang Cosmology

Gino Segrè

VIKING

VIKING
Published by the Penguin Group
Penguin Group (USA) Inc., 375 Hudson Street, New York, New York 10014, U.S.A. · Penguin Group (Canada), 90 Eglinton Avenue East, Suite 700, Toronto, Ontario, Canada M4P 2Y3 (a division of Pearson Penguin Canada Inc.) · Penguin Books Ltd, 80 Strand, London WC2R 0RL, England · Penguin Ireland, 25 St. Stephen's Green, Dublin 2, Ireland (a division of Penguin Books Ltd) · Penguin Books Australia Ltd, 250 Camberwell Road, Camberwell, Victoria 3124, Australia (a division of Pearson Australia Group Pty Ltd) · Penguin Books India Pvt Ltd, 11 Community Centre, Panchsheel Park, New Delhi–110 017, India · Penguin Group (NZ), 67 Apollo Drive, Rosedale, Auckland 0632, New Zealand (a division of Pearson New Zealand Ltd) · Penguin Books (South Africa) (Pty) Ltd, 24 Sturdee Avenue, Rosebank, Johannesburg 2196, South Africa

Penguin Books Ltd, Registered Offices: 80 Strand, London WC2R 0RL, England

First published in 2011 by Viking Penguin, a member of Penguin Group (USA) Inc.

10 9 8 7 6 5 4 3 2 1

Grateful acknowledgment is made for permission to reprint excerpts from the following works:
What Mad Pursuit: A Personal View of Scientific Discovery by Francis Crick. Used by permission of The Perseus Books Group. · *Scientific Correspondence with Bohr, Einstein, Heisenberg, a.o.*, Volume III, by Wolfgang Pauli, edited by K. von Meyenn. Used by permission of Springer Verlag. · *What Is Life?: With "Mind and Matter" and "Autobiographical Sketches"* by Erwin Schrödinger. Copyright © 1967 Cambridge University Press. Reprinted with permission of Cambridge University Press. · *Courageous Hearts: Women and the Anti-Hitler Plot of 1944* by Dorothee von Meding. By permission of Berghahn Books. · *The First Three Minutes* by Steven Weinberg. Used by permission of The Perseus Books Group.

Photograph credits
Insert page 1 (top): Niels Bohr Archive, courtesy AIP Emilio Segrè Visual Archives, Margrethe Bohr Collection; 1 (bottom): AIP Emilio Segrè Visual Archives, Bainbridge Collection; 2 (top): Niels Bohr Archives, Copenhagen; 2 (middle), 3 (bottom), 4 (bottom), 5 (bottom): Courtesy Jonathan Delbrück; 2 (bottom): AIP Emilio Segrè Visual Archives, George Gamow and *Physics Today* Collections; 3 (top), 5 (top): Cold Spring Harbor Laboratory Archives; 4 (top), 6 (top), 7 (bottom): Courtesy Igor Gamow; 6 (bottom): AIP Emilio Segrè Visual Archives, George Gamow Collection; 7 (top): Reni Photos, courtesy AIP Emilio Segrè Visual Archives; 8: James D. Watson Collection, Cold Spring Harbor Laboratory Archives.

LIBRARY OF CONGRESS CATALOGING-IN-PUBLICATION DATA
Segrè, Gino.
Ordinary geniuses : Max Delbrück, George Gamow, and the origins of genomics and big bang cosmology / Gino Segrè.
     p.   cm.
   Includes bibliographical references and index.
   ISBN 978-0-670-02276-2 (hardback)
   1. Delbrück, Max.   2. Molecular biologists—United States—Biography.   3. Gamow, George, 1904–1968.   4. Physicists—United States—Biography.   I. Title.
   QH31.D434S44 2011
   572.8092—dc22
   [B]
   2011009309

Printed in the United States of America
Designed by Carla Bolte

For the **mavericks** of science, young and old.
Enjoy the quest.

Contents

Introduction

I decided to choose myself a corner where nobody was doing anything, so I chose nuclear physics.

George Gamow

Don't do fashionable research.

Max Delbrück

A freshman in my Introductory Physics class recently asked me what I thought were the most exciting areas in science today. She was a very good student, one of the best in the class. She was interested in pursuing a career in basic science but uncertain about which avenue to pursue. I felt she would do well at anything she set her mind to but was hesitant to specify a particular area, so I gave her some general advice, telling her to keep her options open, try working in a few labs, get a summer job doing research, and then see what caught her fancy.

She wanted a little more than generalities, though, and questioned me further. Feeling the need to respond, I told her that my top choices for her were cosmology, the study of the universe; and genomics, the analysis of organisms' genomes. One deals with the world at the macro level, the vast cosmos we are a part of, and the other with the world at the micro level, the smallest aspects of living creatures.

Cosmology has what I consider the greatest unsolved question in the physical sciences, or at least the greatest one that is likely to be addressed successfully in the next two decades: What is our universe made of? All the observed matter, such as stars, planets, and dust, constitutes only 5 percent of what gravitational action tells us is present. We give the remaining 95 percent the names dark energy and dark matter, names chosen because these entities do not seem to emit any recognizable signal. Discovering their identity will surely be a turning point in our view of the universe.

Genomics, like cosmology, has several great unanswered questions. One of them can be succinctly phrased as, how do we decipher the set of instructions for the assembly of living organisms? In the 1970s, when the initial forays into sequencing DNA were made, scientists were surprised to find that most of the information placed in the genetic code did not seem to specify any known function. The analogy to cosmology might have worked better if biologists had chosen to call the seemingly surplus data *dark DNA,* but they opted instead for a more colorful term: *junk DNA.* In recent years, after discovering some of its functions, they have replaced this amusing but seemingly pejorative label with the term *non-coding DNA.* Still, puzzles about its purpose abound. In addition, as gene sequencing becomes ever more powerful and ever less expensive, we are likely to realize the long-awaited revolution in medicine that knowledge of our genetic structure promises.

My student seemed clearly excited about these challenges. I know that by the time she starts her career, new research instruments will be yielding data that cannot be predicted now. After she left my office that day, I found my thoughts drifting from the futures of cosmology and genomics to their pasts, thinking about how they achieved their present status and who the key players in their respective developments were. Doing so led me to consider the intertwined lives of two individuals near and dear to me.

Both were born in the early years of the twentieth century. Trained initially as physicists, they both continued to think of themselves this way even after moving to other fields. They met briefly in the German university town of Göttingen and became reacquainted in Copenhagen, Denmark's capital. They established a friendship there and even worked together for a while before going their separate ways, only to reconnect years later in their new country, the United States.

George Gamow, usually called Geo (pronounced "Joe"), was Russian, and Max Delbrück, invariably known simply as Max, was German. Geo came to be seen as the father of Big Bang cosmology, the first to glimpse how the expanding hot cauldron of the universe's first minutes could form the present world's elements. Max became an iconic figure in the study of bacteriophages, the viruses that provide the simplest means of studying genetic replication. Focusing on these viruses was a key step in uncovering the structure of DNA. Geo and Max can rightly be viewed as founders of modern cosmology and genomics, the two groundbreaking fields I was advising my student to pursue.

Yet, though each of them helped launch a revolution in science, I do not think of them as extraordinary geniuses. Max and Geo are not like the three men who helped steer the quantum mechanics revolution: Wolfgang Pauli, Werner Heisenberg, and Paul Dirac. They did not receive Nobel Prizes for work they accomplished in their twenties, were not awarded prestigious professorships at a tender age, nor were they hailed as leaders in their field before they turned thirty. They also did not, like Francis Crick and James D. Watson, ever write an early paper that changed the direction of science, much less one that reshaped our notions of space and time as Einstein did in 1905. Still, their work has had a lasting influence and has changed modern science.

Pauli, Heisenberg, Dirac, and of course Einstein were extraordinary geniuses, whereas Max and Geo were only ordinary geniuses, smarter and more imaginative than you and me, but not qualitatively different from us. The famous mathematician Mark Kac made the distinction somewhat differently, comparing ordinary geniuses with magicians. These are individuals whose inventions are so ingenious it is hard to see how any mortal could have imagined them. Theirs are not, however, necessarily the most influential creations. In the right circumstances, through good judgment, perseverance, character, and, we must not forget, luck, even ordinary geniuses can lead a revolution in science. To their enduring credit, this is exactly what Max and Geo did.

Part of the reason Max and Geo are so near and dear to me is the greatness of what they achieved, but there are other reasons why I find them so attractive. I certainly appreciate their consistent effort to look at the big picture, to think big thoughts and remain undeterred even when proved wrong. A remark made by Alfred Hershey, a biologist who shared the Nobel Prize with Max, stuck with me. He described how Max *"kept his eye on the big questions, even before they could be put into words. This is something few scientists can do."* These words could apply equally well to Geo, a man who always seemed to be years ahead of his contemporaries. Being so requires formidable intuition, certainly a quality I admire in both Max and Geo, but there is something they possessed that I value even more highly: intellectual courage. Each had it in abundance, combining it with quirkiness in thought and behavior that frequently led them off the beaten path.

Their story would be less interesting had they not also been such lovers of fun. Geo always attempted to transform any gathering into revelry, and his practical jokes became the stuff of legend. On the surface Max appeared quieter, but as his close friend and collaborator Salvador Luria said, *"Max's attitude was Dionysian: a group of students became almost immediately a party and his path through generations of them was like a moveable feast."*

What Luria says about Max's attitude was not limited to students.

I was reassured in my decision to juxtapose in a single story the lives and careers of these two individuals by a recent visit I paid to

Jim Watson, the co-discoverer of DNA, whom I have known since the early 1960s. His beautiful corner office in Cold Spring Harbor, the biology research laboratory he has done so much to develop, is decorated with interesting art, bookshelves, and portraits of Jefferson, Lincoln, and Gregor Mendel. A collage of Crick and Watson by Dalí hangs on the wall, centered directly behind Jim's desk. That day, two photographs caught my eye immediately. One was of a smiling Geo and the other of a youthful Max. In a rambling exchange that lasted several hours, Jim wanted to know more about Max's physics work, and I wanted to discuss his influence on biology; in the end we managed to do both. How much Max had meant to Jim was underscored for me when I later discovered the words Watson spoke at a memorial service held at the California Institute of Technology (Caltech) for Max:

> I still cannot accept the fact that Max is not here and worry that my words will not please him. I want badly to say what I never had the courage to reveal—save now for my wife and children—that Max meant more to me than anyone else.

As for Geo, who is one of the subjects of Watson's amusing book *Genes, Girls, and Gamow,* Jim was reluctant to agree with my characterization of him as an ordinary genius. He believes Geo should be given more credit for being *"so very often a step ahead of everybody."*

Max's and Geo's lives, caught up in the twentieth century's great scientific and political dramas, provide an interesting illustration of what might have been. Max, a scion of one of Prussia's great families, was the resident theoretical physicist in the Berlin nuclear physics group formed by Lise Meitner, who was the first to imagine the possibility of nuclear fission. By the time fission was envisioned, however, Max had realized that his political outspokenness had left him little hope for a scientific career in Fascist Germany. He left his homeland and switched to biology. Had matters been different, his career trajectory might have led him to a Göttingen professorship like the one held by his older cousin and childhood science mentor Karl Friedrich Bonhoeffer. On the other hand, he might instead have met the same end as Karl Friedrich's younger brothers, Dietrich and Klaus Bonhoeffer, Max's contemporaries and childhood friends. One was the well-known Protestant theologian and the other an attorney (married to Max's sister Emilie, or Emmi); the Nazis executed both in the closing days of World War II for their resistance to Hitler.

Geo was even closer than Max to the development of the notion of nuclear fission, for he was the first to think of the nucleus as an oscillating drop of nuclear fluid, the key to imagining the possibility of its splitting into two pieces. Later, Geo could have been a participant in the Manhattan Project's building of the atom bomb, had an early stint as an officer in the Russian Army not counted against him. On the other hand, had he not escaped from Stalin's Russia in 1933, he almost certainly would have suffered arrest in the purges of the late 1930s. Geo's best friend, the great Russian physicist Lev Landau, almost died at that time during a year in prison. The irrepressible Geo would certainly have fared no better.

Max and Geo also shared a feature that requires considerable self-confidence. Gamow put it succinctly in an interview near the end of his life, saying, *"I like the pioneering thing."* He explained that in the late 1920s, when the cream of theoretical physicists, the likes of Pauli, Heisenberg, and Dirac, were attempting to solve the problems of the atom, he followed another path.

I decided to choose myself a corner where nobody was doing anything, so I chose nuclear physics. And in time nuclear physics blew up into a big thing, so I moved to nuclear astronomy, to nuclear astrophysics, to cosmology.

Max, poised in 1932 to become Pauli's physics assistant, opted instead to begin a slow metamorphosis. Deciding to become a biophysicist, he chose a path that demanded years of toiling in obscurity.

Two decades later, when Watson and Crick uncovered the structure of DNA, Max was at the height of his influence, regarded by many as the father figure of the new field that was opening up. And yet he then moved on to something different, the study of a fungus called *Phycomyces*—hardly a step into the spotlight. Many were puzzled, but Max was simply following one of his own oft-quoted aphorisms, *"Don't do fashionable research."*

Rose Bethe, the widow of the great Nobel Prize–winning physicist Hans Bethe, gave me her own view on the two. When I told her I intended to write a book about Max and Geo, both of whom were friends of hers, Rose smiled and quickly said, *"Ah, the two mavericks!"* I, of course, agreed with her.

There is a sort of romance in science similar to what appears in other walks of life. It centers on setting out for a new frontier after the hero or heroine has seemed to achieve his or her goals. We see it in Westerns when the principled gunfighter goes off into the sunset to look for a new town, or in sports when the superstar athlete leaves at the top of her game. We see it in literature, as in Canto XXVI of the *Inferno* of the *Divine Comedy,* where Dante has Ulysses gather his old crew and set sail once again rather than enjoy old age in Ithaca. And we see it in science.

In all these cases the departure carries a price: that of losing the security that comes with being a member of an established community and the support provided by the company of one's colleagues. There is the added risk that the trip may be fruitless or, worse, that a safe return is no longer possible. Still, some individuals have the courage to make it, even feel the need to do so; for them the quest is a necessity, not an option. Most people setting out on such journeys are never heard from again, but part of the romance of any field lies in keeping the dream alive, in not settling for what is familiar and comfortable. This is the story of two extraordinary men who set out on such a voyage and, in the process, led two of the most important science revolutions of the twentieth century.

1 When Max and Geo First Met

I found out at an early age that science is a haven for the timid, the freaks, the misfits. That is more true perhaps for the past than now.

Max, on the atmosphere in Göttingen

It was 1928. Twenty-two-year-old Max Delbrück was sitting at a table in a café in the small central-German university town of Göttingen when a friend pointed out to him a strange young man walking by, a new arrival in town. It was Geo, twenty-four-year-old George Gamow, very tall, very blond, with a big head and myopic eyes that squinted behind thick glasses. Few Russian scientists had been seen in the West since the Revolution, and no students at all. Here was the first, having come to the University of Göttingen for a three-month visit with funds provided by a Russian fellowship. His manner, as Max soon found out, was quite different from that of the physicists he already knew.

In a curiously high-pitched, squeaky voice, somewhat at odds with his imposing figure, Geo made his views known quickly and loudly, despite his having only a meager command of German. Lack of either proper vocabulary or sentence structure was never an obstacle for Geo in expressing himself, which he did often without prompting. He also paid little attention to the traditional regard for rank and status or the customary deference German students showed for a *Herr Professor.* He might have been dismissed in Göttingen as an anomaly, a curious and not entirely welcome product from the new Russia, had he not, in his first week in Göttingen, made a sensational finding: the initial application of quantum mechanics to the study of the atomic nucleus.

By contrast with Geo, Max was at first sight very much the ideal Göttingen student. He was tall and handsome, from a great German family that featured many extremely distinguished academicians. Nevertheless, he seemed to regard with irony the traditional niceties of behavior that marked university life in Göttingen. He was also self-assured in a way that some perceived as arrogance and others as indifference.

Max had enrolled at Göttingen's university two years earlier, intending to become an astronomer, but had soon switched to physics, caught up in the excitement surrounding the emergence of quantum mechanics. He would have liked to get to know the strange Russian better, but there was little time. Their friendship truly began to develop only three years later, at Niels Bohr's Institute in Copenhagen.

Göttingen was a physics mecca in 1928, particularly for those with a strong mathematical preparation. The tradition had been established a century earlier by Karl Friedrich Gauss, perhaps the greatest mathematician of all time, moreover one with a strong interest in physical problems. That tradition was now being carried on by his Göttingen successors, the respective occupants of the chairs of mathematics and theoretical physics, David Hilbert and Max Born, each a towering intellectual figure in his own right. In addition, Hilbert and Born were each entitled to select an assistant to aid in teaching and guiding students, so their own prestige ensured that extraordinary young scientists would occupy those positions. Wolfgang Pauli and Werner Heisenberg had just finished their terms as Born's aides, and John von Neumann, the astonishing Hungarian polymath, had completed months earlier his stay as Hilbert's.

When Max enrolled at the university, students from around the world (for example, the brilliant twenty-two-year-old American J. Robert Oppenheimer) were arriving in the small German town to learn with the masters. Arnold Sommerfeld in Munich could provide an equally solid early training, and Niels Bohr in Copenhagen had a more intuitive approach to physics, but if you were in your early twenties during the crucial 1924–1928 period in the development of quantum mechanics, Göttingen was the place to be. It wasn't necessarily easy, though; the young men and occasional woman were often vying with one another for recognition and success. Max, who seemed to stand aside from the competition, later in life reminisced about what it was like:

> I found out at an early age that science is a haven for the timid, the freaks, the misfits. That is more true perhaps for the past than now. If you were a student in Göttingen in the 1920s and went to the seminar "Structure of Matter" which was under the joint auspices of David Hilbert and Max Born, you could well imagine that you were in a madhouse as you walked in. Every one of the persons there was obviously some kind of a severe case. The least you could do was put on some kind of stutter. Robert Oppenheimer as a graduate student found it expedient to develop a very elegant kind of stutter, the "njum-njum-njum" technique. Thus if you were an oddball, you felt at home.

However, despite being oddballs, neither Max nor Geo felt completely at home in Göttingen's atmosphere. The main reason was that the emphasis on mathematical rigor in solving physics problems was something neither of them was comfortable with. They wanted to find their own approaches. And of course they did, though in Max's case, as we shall see, it took him years to discover one.

2   Max Grows Up

This relatively affluent residential suburb after the war became
almost a ghost town.

Max, on his childhood neighborhood after World War I

Max was born in Berlin on September 4, 1906. His family
belonged to the very highest level of the Prussian bourgeoisie. His
mother's father was surgeon-general of the Prussian armed forces and
her grandfather, Justus von Liebig, was one of the founders of modern chemistry. Max's father's family, equally distinguished, included a
long line of judges, professors, and civil servants.

Max's father, Hans Delbrück, was a history professor at the University of Berlin at a time when important university professors were
quasi deified in the public eye. Earlier he had been a tutor to the
Prussian crown prince and a member of Parliament. In addition to
his professorial duties, the senior Delbrück became a public figure
by editing for almost thirty years the *Preussische Jahrbücher*, a liberal monthly magazine (Max would later compare it to *The Atlantic
Monthly*), for which he regularly wrote a sixteen-page commentary.
The liberal stands he took in the magazine were often controversial,
so much so that the Berlin police chief, apparently insulted by Delbrück's criticism, once challenged him to a duel. Max's father refused
to fight, letting it be known that he thought police chiefs had no business being involved in duels.

Hans Delbrück's position was delicate, since the interpretation of history and, in particular, the history of warfare, his special expertise, mattered a great deal to a Germany that was increasingly eager to establish its claim to power on the world scene. Guided by the country's able and ambitious nineteenth-century leader, Otto von Bismarck, Prussia had vanquished Austria in the war of 1866 and France in 1870. Now at the core of a unified Germany, Berlin had become the capital of a new nation that was emerging as one of the world's great powers.

Bismarck had wanted the city to reflect its recently acquired status by having a boulevard that would rival Paris's Champs-Élysées in beauty and grandeur. The Kurfürstendamm (Berlin's beloved Ku'damm) was the natural choice. Established in the sixteenth century as a trail connecting the ruling Prince Joachim's downtown Unter den Linden palace to his hunting lodge in the western woods outside the city, the street had remained a simple thoroughfare until the second half of the nineteenth century. Having the desired potential to grow in scope, it clearly needed a makeover, however, if it was to reflect the new capital's ambition.

Work on the project began in the early 1880s. The street was widened to span more than 150 feet; cafés, restaurants, and elegant stores sprang up along its path. Prussia's new wealth fueled this change; part came from the country's rapid industrial rise and part from the five billion francs France had been forced to pay in 1870 as war reparations. Villas and elegant apartment buildings reflected this new prosperity, with the Ku'damm now a favorite site for construction. Toward the end of the nineteenth century, Grunewald, an area at the boulevard's western edge, also began to be developed, a reflection of the city's natural expansion.

Max's parents, who until his birth had lived in a city apartment, felt they wanted something more comfortable with the arrival of baby Max, the last of their seven children. In 1907 they moved to this newly established Berlin suburb, a desirable address for well-to-do academicians and successful businessmen. The Delbrücks were drawn there for a number of reasons, more space and greenery being obvious ones. Distance from downtown was no longer a barrier, since the commuter train ran all the way from the center of Berlin to Grunewald. Having friends and relatives in the immediate vicinity, however, was probably the chief motivation for their relocation.

Three families chose to live within a few minutes of one another: the Delbrücks, the Bonhoeffers, and the von Harnacks. They represented the very cream of German intelligentsia and were, in addition, related. Max's mother, Lina Delbrück, and Amelie von Harnack were sisters, and the Bonhoeffers were their cousins. In each case the father was a prominent University of Berlin professor—moreover, one known for his strong moral positions and interests in reform. Karl Bonhoeffer was a neurologist and psychiatrist, a shaper of German mental health programs. Adolf von Harnack, Germany's foremost liberal theologian, had argued for examination of all facets of Christianity, denied the existence of miracles, and emphasized the links between Greek philosophy and early Church writings, all controversial positions in Berlin. Also noted for his organizational skills, by the time Max was born, von Harnack was director of the Royal Library and rector of the university.

The Delbrücks, the Bonhoeffers, and the von Harnacks usually gathered together on Sunday nights for free-ranging discussions, in which the older children participated while the younger ones played. The three families' ideas on education were unusual for the time. A Bonhoeffer family saying was that *"Germans had their backs broken twice in the course of their lives; first at school and then during military service."* As a consequence, the eight Bonhoeffer children were homeschooled. The Delbrücks had their own eccentric views. Once, observing the boys playing chess on the roof by the chimney, Max's sister Emmi heard a neighbor calling out to her father, *" 'Do you know that your sons are sitting on the roof? Isn't that dangerous?' To which my father replied, 'Don't look at them.' My father was convinced that when a child embarked on something dangerous like that without trying to show off, it proved that he had the situation under control."*

Independence was fostered in the three families, but there was also a strong emphasis on moral behavior and on making hard choices, even if they proved unpopular. Emmi remembered her uncle Adolf (von Harnack) telling her as a ten-year-old, *"If you are ever uncertain about which direction to take—right or left—do what you least enjoy doing; it usually turns out to be the right thing."* This lesson stayed with the group of children when, in later years, as adults, they battled Hitler and the Nazi regime instead of going along with it, the way others were doing.

The happy life of the three families came to an end, in 1914–1918, with the tragedies of World War I. Germany was initially swept up in a wave of patriotism. Von Harnack said that it would bring *"one will, one force, one seriousness of purpose,"* directing the *"holy flame of the fatherland"* against *"everything selfish, petty and common,"* but by 1915 the mood had changed and the Grunewald group was now trying to moderate the right-wing German expansionist policy sweeping the country. When a so-called intellectuals declaration advocating annexation of neighboring territories was issued, Hans Delbrück formulated a counter-declaration rejecting such aims and favoring internal reform. Fifty Berlin professors signed the first declaration, while only fifteen signed Delbrück's, but among these fifteen were von Harnack, Planck, and Einstein.

The terrible loss of young men's lives in the trenches, the starvation that prevailed during the war, and the ensuing unrest and alienation hit Germany especially hard. Max's brother Waldemar had been killed, as had the Bonhoeffers' son Walter. Max's mother kept on her desk a large circle of photographs of friends and relatives, all of whom had died in the war. As for Max, he remembered the dramatic change to the setting of his happy childhood, how the *"relatively affluent residential suburb after the war became almost a ghost town."*

Perhaps seeking a separate identity, or simply attempting to distance himself from the swirling crosscurrents of politics, history, and economics, Max pursued a passion he had developed for astronomy, a field that was entirely new to the family. It may also have been a first manifestation of his desire to do *"the pioneering thing."* He acquired a small telescope, which he mounted on a stand placed on a little balcony adjacent to his parents' bedroom. When sightings had to be taken at 2:00 a.m., which was not infrequent, the teenage Max would set an alarm (he remembered it being very loud), traipse through his parents' bedroom, spend an hour on the balcony, and then go back to bed.

Seeing his interest in science, Max's older cousin Karl Friedrich Bonhoeffer, later a distinguished professor of biochemistry at the University of Göttingen, took him under his wing. Seven years Max's senior, Karl Friedrich gave him books to read and encouraged his pursuits. Max hung in his room a picture of Kepler, the great astronomer who more than three centuries earlier had concluded that planetary orbits were ellipses, not circles, and who then established the three laws of their motion that laid the groundwork for Newton's grand synthesis.

Max gave a valedictory address about Kepler at his high school graduation and, the following fall, set off to study astronomy at the University of Tübingen. It was customary in those days for German students to move from one university to another over the course of their undergraduate careers, so he soon transferred to Bonn, then went on to Berlin, and in 1926 arrived in Göttingen, intent on becoming an astronomer. His interest fell on the phenomenon of novae, stars that over the course of days increase in brightness by several orders of magnitude. But he wanted to do more than chronicle their curious behavior. He was looking for an explanation of the rapid changes in their luminosity, and this pointed him to the unknowns in stellar energy generation. He then saw that even the source of the sun's output was still a mystery, so there seemed little hope of gaining the kind of fundamental insights he was looking for. These would come only a decade later, once a better understanding of atomic and nuclear physics (a field that would be pioneered by Max's future friend Geo) had become available and could be applied to stellar phenomena.

Still, Max's love of astronomy remained, and many happy hours were spent in his later life observing and recording the movement of stars in the night sky. This would, however, never be more than a leisure pursuit in those years.

As Max was becoming conscious of the obstacles before him in astronomy, he began hearing of the growing excitement in physics. In 1925, while he was still in Berlin, he learned that a twenty-four-year-old from Munich had achieved a breakthrough in atomic physics/quantum theory. Later that year Max went to a lecture on the subject by this young man, whose name was Werner Heisenberg. Though Max understood very little of what was said, he was impressed by the attention of the standing-room-only crowd and by overhearing beforehand Einstein tell some of Berlin's other greats that this was a very important development.

In 1911 the great experimental physicist Ernest Rutherford had shown that the atom was composed of a minuscule but very massive nucleus surrounded by orbiting electrons. Two year later Niels Bohr had proposed a model of the atom that incorporated Max Planck's quantum theory postulate that energy was emitted or absorbed in so-called quanta, discrete packets. Showing how the electrons rotated around the nucleus, Bohr's model, which resembled a miniature solar system, was in amazing agreement with data obtained from hydrogen, the simplest of all atoms. It encountered serious difficulties, however, in treating more complicated atoms and was plagued by apparent inconsistencies in attempting to bridge the connection between the quantum and classical descriptions of the submicroscopic world. By 1925 it was becoming increasingly clear that a radical departure from Bohr's model would be needed. At that time Heisenberg and, independently, Erwin Schrödinger, provided a way to proceed. Quantum mechanics was born.

The ensuing shift was comparable to that between studying motion before and after Newton's formulation of the laws of mechanics and his analysis of the gravitational force. Kepler had provided the clues with his three laws, but after Newton, planetary motion could be calculated in exquisite detail, and any deviations from that motion led to spectacular predictions, such as the existence of not-yet-observed planets. Newton's laws were a watershed.

So, too, with quantum mechanics. The excitement in the physics world was enormous after the discovery of quantum mechanics, as classes of problems suddenly seemed open to new approaches with the tools it provided. Atomic physics was of course changed, but so were notions of molecular binding, magnetism, electrical conductivity, radiation, and many other phenomena. The problems encountered in looking for solutions were often formidable, but by 1929 most physicists would subscribe to the statement Dirac made that year:

The underlying physical laws necessary for the mathematical treatment of a large part of physics and the whole of chemistry are thus completely known, and the difficulty lies only in the fact that application of these laws leads to equations that are too complex to be solved.

Quantum mechanics was nothing less than a revolution in science.

This was the new world of physics that Max Delbrück was seeing emerge in Göttingen. He could not help but feel the excitement of the young mathematicians and physicists pursuing the subject's implications. The first one with whom he interacted closely was Eugene Wigner, one of the half dozen or so extraordinary young scientists emerging from Budapest's Jewish community (for example, Leo Szilard and John von Neumann). At the time working as the mathematician David Hilbert's assistant, Wigner was beginning to develop mathematical techniques to deal with the complicated symmetry properties that determined the quantum mechanical behavior of many particle assemblages. These techniques would eventually prove enormously influential, but in 1927 the subject was in its infancy, and its leading exponent, at twenty-six, was still a very young man.

Max tried to understand Wigner's work but lacked the necessary background. However, he discovered an old book by a Swiss mathematician that had been republished in 1927 as part of a series on mathematics and its applications. It delved into an examination of pattern regularities going as far back as Egyptian tombs. Along the way, the book discussed some of the fundamentals of group theory, the branch of mathematics Wigner was extending. Max found there the tools he needed to understand the new work and even to complete a few proofs his Hungarian friend had omitted. When Max brought the matter to Wigner's attention, Wigner advised him to publish what he had uncovered. At first Max declined, maintaining that what he had done was too trivial, but Wigner was insistent—rightly so, since the existence of the resulting paper made others in Göttingen take note that Max was capable of undertaking independent research.

It wasn't a particularly important piece of work, nor was it anything that Max later pursued, but it showed he was capable of independently moving off in a new direction and acquiring the tools to pursue a question that interested him. This was a skill he would later demonstrate far more significantly.

3 Geo Grows Up

In more than twenty years together, Geo had never been happier
than when perpetuating a practical joke.

Geo's first wife, Rho, in Watson, Genes, Girls, and Gamow

Georgiy Antonovich Gamov (the Russian form of his sur-
name) was born in Odessa on March 4, 1904; he was an only child.
The surgeon, called in the middle of the night, performed an emergency Caesarean section in the Gamow home with a maid holding a
kerosene lamp aloft. The instruments the surgeon used were steril-
ized in the Gamow kitchen, and the delivery, not an easy one, took
place on the writing desk in Geo's father's study. Geo claimed that his
writing of books as an adult was a consequence of having been surrounded by them at birth.

Odessa was quite unlike other Russian cities. In 1794, in order
to consolidate the Russian hold on the Black Sea, Catherine the
Great, then Russia's tsarina, had ordered that this little trading village, recently captured from the Turks, be built into a city. By the time
Geo was born, it was the fourth-largest municipality in the Russian
Empire after Moscow, St. Petersburg, and Warsaw. It had become a
major port of trade, where, Pushkin once quipped, it was cheaper to
drink wine than water. Cosmopolitan in outlook, it was also Russia's chief harbor for foreign commerce, a city where more than fifty
languages were spoken, where the first published newspaper was in

French, where street signs were sometimes in Italian, and where one might easily encounter Greeks, Japanese, Chinese, and Indians. And then there were the Jews, a third or more of the city's population, drawn there from throughout Russia by commerce and the promise of a better life. Often settling in the ghetto of Moldavanka, a neighborhood immortalized in Isaac Babel's *Odessa Tales,* they constituted a rich subculture.

Geo's paternal grandfather was a colonel in the Russian Army. Three of his grandfather's four sons, having followed him in military careers, were killed in the 1905 Russo-Japanese War. The fourth, Geo's father, Anton, studied Russian literature at the University of Odessa and then began teaching Russian at a preparatory school. Geo's maternal grandfather was the chief cleric of Odessa's cathedral, a metropolitan in the Russian Orthodox Church. This meant that both sides of the family were well established, if not rich.

Wealth in the Russian Empire was mainly in the hands of the nobility, owners of most of the land. Russia, though beginning to industrialize, remained a largely agrarian country farmed by serfs, whose condition was little better than slavery until their emancipation in 1861. Even after that, however, they found it hard to manage more than a subsistence living. Since conditions were not much better in the urban centers, pressure grew, with city workers joining rural serfs in pressing for reform. When the Russo-Japanese War, fought for what many regarded as questionable aims, ended in a humiliating defeat for the Russians, revolution broke out, with particularly fierce battles in Gamow's hometown.

One such incident was immortalized in Sergei Eisenstein's movie *The Battleship Potemkin*. The *Potemkin*'s crew revolted in 1905 and, after taking control of the ship, sailed into Odessa's harbor, to cheering crowds. Tsarist forces fought back, firing on those assembled. The ensuing slaughter on the Odessa Steps famously depicted in the film never took place, but many of the city's citizens were killed in urban battles that year. The revolt was eventually quashed, but Bolshevik forces regarded it as an important precursor to the 1917 Revolution, in which they gained control over Russia.

A period of relative peace and well-being ensued for the Gamows, but in 1913 tragedy struck the family: Geo's mother died. What followed was an unimaginably hard decade for the nearsighted, sensitive young Geo. World War I started the following year, and Odessa, which had once sported magnificent acacia trees, was denuded, the trees cut down in a city desperate for heating materials. Starvation was rampant, with little or no food coming in from the countryside. Even drinking water was often unavailable, because of the breakdown of machinery and absence of fuel to operate the pumps that brought water in from the nearby Dniester River.

After the overthrow of the tsarist regime in early 1917 and the Bolshevik takeover in November 1917, another turbulent phase in Russia's history began. Ukraine, the Russian territory that included Odessa, was particularly hard-hit, quickly turning into a strategic pawn in the ensuing Russian Civil War. Occupied in succession by Bolsheviks, Germans, Ukrainian socialists, French forces, Bolsheviks again, and the White Russian Army, it finally joined other Soviet republics when the Union of Soviet Socialist Republics, or USSR, was formed in 1922. Peace followed, but prosperity did not. From the cumulative effects of World War I, the ensuing civil war, and disastrous revolutionary policies, Russia's economy had collapsed by 1921.

During all these vicissitudes, Geo was attempting to gain an education. His first memory related to science was from age six, when he observed Halley's Comet from the roof of his house with a small telescope his father had bought for him. However, the experiment that persuaded him to become a scientist, he claimed, was performed with a microscope. Being the grandson of Odessa's metropolitan, he was expected to excel in religious studies, then obligatory in school. One day he decided to test whether the small piece of bread dipped into red wine and placed on his tongue during Communion in the Russian Orthodox Church had been transubstantiated into the flesh and blood of Jesus Christ. Instead of swallowing the bread, he kept it in his cheek and ran home to scrutinize it under the microscope's lens. He found it was quite unlike the piece of skin he scraped from his own finger, but entirely similar to the piece of bread dipped in wine that he had prepared beforehand, thus confirming his suspicion that the bread had remained unchanged.

Science was not Geo's only interest. He also began to learn reams of Russian poetry, with Pushkin a special favorite; in later life he would recite them at the drop of a hat. In 1922, at age eighteen, he enrolled in the University of Odessa as a physics student, but the physics classes were all canceled because no supplies were available. As for mathematics lectures, these were often held in the dark, since electricity had been cut off. Deeming the situation hopeless, Geo decided to transfer to a university in St. Petersburg (which was known as Petrograd from 1914 to 1924, when its name was changed to Leningrad; it is now again called St. Petersburg), having heard that things were better there. Geo's father sold most of the remaining family silver to give his son some money for the trip.

Arriving in the former Russian capital, Geo was fortunate to find a job that gave him enough to live on and left him time to study. Working for a meteorologist, an old friend of his father's, he was asked to record weather measurements at outdoor sites for the city's Forestry Institute. These had to be taken seven days a week for twenty minutes at 6:00 a.m., noon, and 6:00 p.m. It was a rigorous schedule, but as Geo observed, he had to work for only sixty minutes a day.

In 1924, while still a student, he obtained a new, more prestigious, and better-paying job, that of teaching physics to military cadets in the Artillery School of the Red October. So he could obtain a salary on a scale commensurate with his duties, he was made a field army colonel and given a beautiful uniform, black riding boots with spurs, and a conical hat with a red star. In later years he regretted having lost the photograph of himself, at age twenty, resplendently dressed in this regalia.

A year later the regular physics teacher, who had been away on a leave of absence, returned, and Geo's career in the Red Army ended—but it was not completely forgotten. Many years later and almost ten thousand miles away, while Geo was swimming off the shore at a California resort, he saw two men dressed in business suits running up and down the beach calling his name. One of them was from the FBI and the other from the security division of the AEC (Atomic Energy Commission). Without telling him why, they flew him to Los Alamos on an emergency basis. On arrival at the New Mexico laboratory, where he was acting as a consultant, Geo was informed that Senator Joseph McCarthy was planning to announce the next day that the AEC had recruited a former Red Army colonel without having performed an adequate security check on him. Once Geo explained to the head of laboratory security, however, what his true role in the Soviet Army had been, smiles reappeared in the security division.

Gamow tells this story and others like it in his brief autobiography, *My World Line*. Some humorous anecdotes were doubtlessly embellished, though perhaps not as much as one might think, for many of Geo's adventures were often quite out of the ordinary, and the times he was living through certainly were. There is not a hint of self-pity in the book, but one can only imagine the toll taken on him by those eleven years of his youth between the death of his mother in 1913 and his 1924 arrival in Leningrad. Yet in later life he presented an almost unrelenting cheerfulness. As Rho, his first wife, said, *"In more than twenty years together, Geo had never been happier than when perpetuating a practical joke."* (Such jokes were, as we shall see, usually directed at colleagues who took themselves too seriously.) One might conjecture that his humor, often well lubricated with alcohol, was a defense against blacker thoughts, but before jumping to that conclusion, we would do well to remember that wit and joking were pervasive in his birthplace of Odessa, a city where the rich veins of Yiddish and Russian humor traditionally enriched each other.

After finishing his course work at the university, Geo was admitted to a program that enabled him to acquire the equivalent of a modern-day Ph.D. The subject that at first interested him most was the study of gravity as developed by Einstein in his general theory of relativity. Geo was fortunate that the faculty in Leningrad included a brilliant young mathematician, Alexander Friedmann, an expert on the subject, but Friedmann died not long after Geo's arrival.

Geo had been hoping to work on problems of cosmology under Friedmann's direction, but he now had to look elsewhere. Having gotten into the habit of perusing whatever foreign physics journals were available in the library, he remembered encountering quantum mechanics through reading Schrödinger's first paper on the subject in a 1926 issue of the *Annalen der Physik*. Nobody on the university's physics faculty appeared very interested in the subject, but Geo quickly grasped that it presented an entirely new way of studying atomic physics problems and was likely to lead to new and interesting developments. Fortunately, so did two other Leningrad physics students. Together, the three men learned quantum mechanics without guidance from the faculty, developing their own style, one that relied much more on intuition than the more formal approach favored in Göttingen. In the process, and as Max had done in Göttingen, the three learned independence, too, and decided they would not be intimidated by anybody.

Lev Davidovich Landau, Dmitri Dmitriyevich Ivanenko, and Georgiy Antonovich Gamov—sometimes known as Dau, Dim, and Geo—studied together, worked together, played tennis with one another, organized parties, wrote humorous verse, and made fun of their elders who were unwilling or unable to gain expertise in the new quantum mechanics. The three of them came to be admired by the younger students for their ability, their intellectual daring, and their obvious joie de vivre. Each of the three later had an extraordinary career, none more so than the incredibly precocious Dau, who had arrived in Leningrad in 1924 as a sixteen-year-old college graduate (he later said he couldn't remember a time when he had not understood calculus).

Joining Geo a few years later in Copenhagen, he went on to become one of the world's greatest theoretical physicists.

Nevertheless, the absence of guidance by elders and of contact with physicists from abroad made intellectual life hard for the Three Musketeers, as they were also known. Fortunately for Geo, a professor at the university recommended in 1928 that he be awarded a three-month fellowship for study at the University of Göttingen. The fellowship was granted, and in June, with Dim, Dau, and other friends cheering at the dock, Geo took a boat from Leningrad to the German port of Swinemünde, and from there proceeded by train to Göttingen.

4   Göttingen and Copenhagen

The explanation did not appeal to me at all, and before I closed the
magazine, I knew what actually happens in this case.

Geo, on an article by Ernest Rutherford about nuclear decay

Almost immediately after arriving from Russia in the small
German university town, Geo went to the library. With an omnivo-
rous interest in physics, he was anxious to read journals, many not
available in his native country, in order to discover what were the new
developments in the subject he loved. One of the first articles he came
across contained a suggestion by Ernest Rutherford for how nuclei,
the massive tiny cores of atoms, decay by discharging some of their
constituents. Rutherford, not a man whose views were taken lightly,
was then probably the world's leading experimental physicist. He was
the Cavendish professor at Cambridge University, the winner of the
Nobel Prize in Chemistry for his work on radioactive decays, the president of the Royal Society, and the discoverer of the atomic nucleus.

Geo read the article quickly to see what the great man was proposing. Thirty years earlier Rutherford had been the first to show that
emissions from radioactive substances could be separated into two
kinds of rays, to which he gave the names alpha and beta. The latter
were composed of electrons; the former were alpha particles, appearing to be ionized (stripped of their electrons) helium atoms. Ten years
later he had shown this was true. Still a little later, aiming highly

energetic alpha particles at a thin foil of gold, Rutherford and his student Ernest Marsden had been stunned to see the projectiles head directly back at them. He famously described this sight as the most incredible one of his life; it was, he said, *"as if you fired a 15-inch shell at a piece of tissue paper and it came back and hit you."* He had discovered the atom's nucleus; in doing so, he had also paved the way for Bohr's model of the atom and all of later atomic physics.

It did not end there. Rutherford continued to be fascinated by alpha particles; now, in 1928, he was reporting on further experiments in which they were scattered by heavy nuclei. At the same time, he was also trying to understand how it was that nuclei would sometimes spontaneously emit alpha particles. This was the very question that had started him out on his journey thirty years earlier. Rutherford was not a theoretical physicist and not particularly interested in sophisticated models. He was known to quip that he wasn't interested in theories he couldn't explain to a barmaid. The model he was proposing for alpha particle emission by nuclei was therefore necessarily little more than an intuitive picture. It involved electrons attaching themselves to alpha particles inside the heavy nucleus, steering those particles to the outside, and then detaching themselves and returning to their origin.

Reading the paper, Geo had a sudden flash of intuition. As he says in his autobiography, *"The explanation did not appeal to me at all, and before I closed the magazine, I knew what actually happens in this case."* A young physicist with less self-confidence might have been hesitant about disagreeing with such a towering figure, but lack of self-confidence was never Geo's problem. He had realized that recent insights into chpter provided by the development of quantum mechanics, which I will describe in the next chapter, provided a way to calculate the probability of a nucleus emitting an alpha particle.

A few weeks of work confirmed for Geo what he had previously only intuited, and he was ready to present his conclusions at the university's weekly physics colloquium. Max Born, the gathering's presiding figure, thought Geo's result was extremely interesting but criticized it on grounds that the techniques Geo had used lacked mathematical rigor and seemed to rely on insight and analogies rather than the clear set of rules usually employed in quantum mechanical calculations. Gamow's response was contrarian; why was Born making such a big fuss about subtleties? Geo had explained several long-standing puzzles in nuclear decays. Wasn't that good enough?

Born's formal approach was at odds with the more intuitive style Geo employed. Born soon managed to obtain a derivation of Geo's formula that satisfied his strict criteria, but Geo's method for calculating nuclear decay probabilities is the one used today. Theirs was a clash of styles, a question of how mathematically rigorous a theoretical physics derivation needs to be. It isn't easy to change the way one approaches a theoretical physics problem, to exchange tools one has developed and prefers to employ—in other words, to alter one's style. We think of style as identifiable in the arts, but it can and often is equally distinctive in the sciences. Success often becomes a question of whether the important problems of the day are solvable in the way an individual likes to think of them. At this point you may object, rightly, that the artist creates his or her own world, while the scientist is only attempting to describe what nature has laid down. But would Brunelleschi be famous if Florence had not wanted a great dome for its cathedral, or would Palladio be hailed as the forefather of modern architecture if the great Venetian families had not built countryside villas during his lifetime? Their skills, great as they were, suited the problems of their time.

Max Born probably did not fully appreciate the young Russian, but fortunately, Geo soon met Niels Bohr, a great theoretical physicist who did. Bohr had felt on occasion the same sort of criticism Geo had experienced, even hearing his correspondence principle, the attempt to bridge the gap between quantum and classical theories, described as *"a somewhat mystical magical wand, which did not act outside Copenhagen."* These are words that would never be used to describe any results obtained by the meticulous Born.

Niels Bohr grew up in Denmark, where he was born in 1885. In 1911, directly after receiving his doctorate, he went off to England for a year's study. A few months in Manchester with Ernest Rutherford were a turning point for him. He and the distinguished experimentalist became life-long friends, interspersing visits together with frequent correspondence. Moreover, the informal and collegial atmosphere Bohr found in Manchester inspired him to create something similar for theoretical physics in his native Copenhagen.

With fresh memories of Manchester, Bohr had begun right away to think of building in Denmark an institute for theoretical physicists who had already mastered the essentials of their craft. Arriving in Copenhagen, they would live, work, and play together. The idea was always to have at least six or seven young physicists in residence, with others passing through, perhaps staying for a few days or weeks. The longer-term residences might last months or even years, depending on circumstances and funding; it was all meant to be fluid. It took Bohr several years of hard work to raise the necessary resources, but his dream was finally realized in 1921. In March of that year the Institut for Teoretisk Fysik formally opened. At the opening ceremonies, Bohr delivered an address stating that the Institute's main theme would be *"The task of having to introduce a constantly renewed number of young people into the results and methods of science."*

The Institute's main door was reached by climbing a few steps. Once inside, the lecture hall was on the right and a small library with desks and chairs on the left. There was a cafeteria, where good black bread, cheese, and coffee were always available. The growing Bohr family, Niels, Margrethe, and their three sons, lived on the top two floors. In 1924 the University of Copenhagen formally relieved Bohr of his teaching duties, thus freeing him to devote all his time to the Institute. The family by now included five sons, the last having been born in March of that year. The old building was no longer large enough, so construction began on two adjacent buildings, one intended for experimental research and the other as a residence for the Bohrs. The space once used by the family in the original building was converted to offices and a small apartment for visitors. All of this was ready by the fall of 1926; Werner Heisenberg was the first visitor to use the apartment. Two years after that Geo would be one of the mainstays at Bohr's Institute.

5 Particle or Wave?

Build me a one-million-electron-volt accelerator; we will crack the
lithium nucleus without any trouble.

Ernest Rutherford, addressing John Cockcroft and Ernest Walton

In order to understand Geo's achievement in 1928 and the
roots of Max Delbrück's eventual move from physics to biology, we
will need to take a step backward and explore some of the development of quantum mechanics, emphasizing again how this was the
great revolution in twentieth-century physics, both for the novelty
of the ideas it introduced (such as the uncertainty principle) and for
the importance of the applications it led to. Our modern world of
semiconductors, lasers, and MRIs flows directly from studies of the
principles of quantum mechanics.

Bohr's 1913 model of electrons circling a central nucleus had been
extraordinarily successful in explaining the features of the hydrogen
atom, but it ran into difficulties when attempts were made to apply it
to larger, more complicated atoms or to use it to understand the features of molecule formation and binding. By 1925 a consensus was
emerging that a radical reframing of the model would be necessary.
Such a framing then appeared, surprisingly, in two forms. Heisenberg
proposed one in the summer of 1925 and Erwin Schrödinger the other
in the winter of that same year. Initially the two went respectively by
the names matrix mechanics and wave mechanics. Schrödinger soon

showed the two to be essentially identical in their predictions; the resulting theory was now simply called quantum mechanics.

Even though both theories explained successfully a series of problems that had plagued the Bohr model, Heisenberg and Schrödinger were at loggerheads on questions of interpretation. Schrödinger retained the notion of electron orbits, while Heisenberg claimed that these were meaningless; according to him, the theory had to be based entirely on observable quantities, and orbits were not observable. Heisenberg's formulation depended on using the mathematical formalism of matrix theory, with restrictions imposed on how the matrices were to be used. Schrödinger found this approach opaque, counterintuitive, and almost useless for practical calculations, because of its absence of a simple visualization.

Most leading theoretical physicists initially sided with Schrödinger, but Bohr did not; he thought some essential features in the theory's interpretation were still missing. In May 1926, having spent the previous year in Göttingen, Heisenberg came to Copenhagen to see if he and Bohr could clarify matters. Over the course of the following year, with some helpful criticism from Wolfgang Pauli, they developed what is now known as the Copenhagen interpretation of quantum mechanics. Though not accepted by everybody, most significantly not by Einstein, it has withstood all tests of its validity for the past eighty-plus years.

Much of the difficulty that had plagued physicists for the previous twenty-five years had centered on the difference between waves and particles. Scientists had thought these were very different kinds of objects, but the distinction in the subatomic world began to blur after Planck's introduction of the notion of an energy *quantum*. Planck first glimpsed this idea on a Sunday evening in October 1900, while working in the study of his Grunewald home, just around the corner from where the Delbrücks would settle seven years later. In fact, Max Delbrück remembered stealing cherries as a boy from the adjacent Planck garden.

Planck's suggestion that light and other forms of electromagnetic radiation appear in packets, or quanta, seemed almost a throwback to Isaac Newton's proposal that light beams were composed of very small particles. In the early 1800s this picture had been disproved, replaced by wave theory, according to which the amount of energy carried by a light beam varies continuously.

Planck did not challenge wave theory head-on, but five years later Einstein did, asserting that quanta could move through space, scatter, or be reflected, acting in every way like particles. By doing so, he explained the "photoelectric effect," a long-standing puzzle over how an electrical current could be emitted by a surface on which radiation was shone. Einstein's proposal left physics in a quandary. Was electromagnetic radiation indeed composed of waves, or could it sometimes behave like a particle? The situation became even stranger in 1923, after a French nobleman, Prince Louis de Broglie, suggested that if a light wave could behave like a particle, wasn't it natural to think of particles and electrons in particular as waves? Admittedly it was bizarre to imagine that particles were waves and waves were particles, but it also seemed correct. What did this all mean?

It was clear that Bohr's old picture of an atom as composed of electrons moving in prescribed orbits about a nucleus was wrong, but de Broglie's interpretation of the electron as a wave was also flawed. Schrödinger attempted to keep the best features of both by having the electron be a particle with motion dictated by a so-called guiding wave: hence the term *wave mechanics* for his formulation of quantum mechanics. But Bohr and Heisenberg maintained that the either-or dichotomy of particle versus wave was wrong. They came to believe that both particle and wave appearances were simply the results of experiments for detecting a moving object. Depending on how the measuring apparatus was designed, either form could be observed. Only one thing was impossible: both particle and wave could not be detected simultaneously. Why not? This was where the uncertainty principle came into play by placing limits on the accuracy of measurements.*

Geo's 1928 theory of how nuclei decay by emitting alpha particles was a brilliant application of envisioning objects as *both* waves and particles, though not simultaneously. In the Göttingen library, he had realized that although the strong forces holding alpha particles inside the nucleus appear to act as an impenetrable barrier, they should be envisioned as constraining but not imprisoning those particles. In their waveform, alpha particles could surmount or, rather, tunnel directly through the obstacle and then reappear on the other side as alpha particles, admittedly with small probability (see figure 1). A fly cannot pass through a closed window, and yet Geo was proposing the seemingly miraculous appearance of that fly on the other side. A distinguished British physicist put it more colorfully while introducing Gamow at a Royal Society talk a year later: *"Anyone present in this room has a finite chance of leaving it without opening the door, or, of course, without being thrown out through the window."*

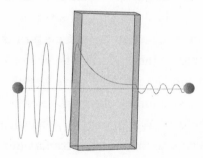

FIGURE 1
Tunneling through a barrier: Diminution of wave amplitude represents diminishing probability that a particle striking on the left will reappear on the right.

*The uncertainty principle maintains that there is a limit to how precisely one can simultaneously measure the position and the velocity (technically, the momentum) of an object. As an example of its applicability, one can see that the determination of an electron orbit breaks down because specifying it would require simultaneously knowing both an electron's position (location of orbit) and its velocity (electron motion along the orbit).

Geo had realized that, with a little work, he could calculate the probability of alpha particles/waves emerging from inside the nucleus. It was the first application of quantum mechanics to nuclear physics, a major coup for Geo. It marked him as a young scientist to be watched, perhaps not in the same league as the slightly older Heisenberg, Pauli, and Dirac, but nevertheless someone possessing rare powers of intuition and capable of obtaining very interesting new results.

The phenomenon of burrowing through seemingly impenetrable walls was also seen to be a much more general one, with applications to many other areas in physics and chemistry. It was soon given the name quantum tunneling.

After three months in Göttingen, Geo had reached a crossroads. His fellowship funds were exhausted, and it was time for him to return to Leningrad. He decided to reroute his ticket home in order to spend one day in Copenhagen, hoping to meet some of the young physicists there and perhaps even talk to Bohr. Arriving at the Institute, he was informed by the secretary that he might have to wait several days to speak to the director, but when he told her he had only one day in Copenhagen, she arranged for him to meet Bohr that very afternoon. Their conversation that day quickly turned to Geo's calculation of nuclear decay by the emission of alpha particles. Geo's knowledge of German, his common language with Bohr, was still rather fragmentary, but the equations spoke for themselves. Quickly realizing that Geo had solved a very important problem and had unusual imaginative power, Bohr said to him, "*My secretary told me you have only enough money to stay here for a day. If I arrange for you a Carlsberg Fellowship at the Royal Danish Academy of Sciences, would you stay here for one year?*" With a sigh of relief, the reply was an immediate, resounding yes.

Believing that Rutherford should also learn of Geo's results on alpha decay, Bohr organized a trip for Geo to visit Cambridge, sending a tactful letter along to his old friend explaining why Gamow's version of events was more believable than Rutherford's. The doyen of experimental particle physics was not fond of complicated mathematics and was often suspicious of theorists, but he liked the freewheeling young Russian and was impressed by the fit to his own data that Geo's model presented.

Once the fellowship supported by the Carlsberg Brewery expired, Geo moved to Cambridge with a Rockefeller Foundation fellowship. He cut a colorful figure riding around the British landscape on his secondhand motorcycle, dressed in his newly acquired knickers and Argyle socks. His old Leningrad friend Lev Landau, the physics prodigy with the wild hair and the lightning-quick mind, joined him in the summer of 1930. The twenty-two-year-old Landau was by now a seasoned physicist. Quickly making plans, the two of them took off for an extensive tour of Scotland, Gamow in front and Landau on the back of the motorcycle. One can only imagine what Scottish villagers thought of these two arriving in town babbling away in Russian about physics.

In the meantime, Gamow continued to focus on the problem of quantum tunneling, as did a Göttingen friend of his, Fritz Houtermans. They independently began to turn the problem around. If an alpha particle or, for that matter, a proton can tunnel out of a nucleus, under what circumstances might one be able to tunnel in? Either a proton or an alpha particle is electrically positive, as is the nucleus; there is a repulsion between them, creating an effective barrier. But if a proton was aimed at a nucleus and penetrated that barrier, it would enter a zone where the strong attractive forces that hold nuclear constituents together dominate. Those forces would take over and might even trap the projectile. If this was the case, the nucleus would surely be altered and perhaps even disintegrate.

Ernest Rutherford was very interested in such a possibility. He asked Geo how likely this was. When an encouraging answer was shot back at him, he summoned two young Cambridge experimental-ists, John Cockcroft and Ernest Walton, telling them in no uncertain terms, *"Build me a one-million-electron-volt accelerator; we will crack the lithium nucleus without any trouble."* With Rutherford and Geo urging them on, Cockcroft and Walton built the necessary appara-tus and performed the experiment in 1932. It earned them the 1951 Nobel Prize in Physics.

Ranging widely in their thinking about science, Geo and his Göt-tingen friend Houtermans now began to consider if there might be situations in which a proton would acquire enough energy to pen-etrate inside a nucleus without the need of an accelerator. Such a scenario wouldn't occur on earth, but perhaps it could occur in the core of stars, where pressures were enormous and temperatures rose to millions of degrees. Might this or the even simpler fusion of pro-tons to form alpha particles be an energy source for stellar cores?

In the nineteenth century, known power sources such as oil or coal had been shown to be grossly inadequate in sustaining a solar lifetime long enough to heat the earth during the millions of years necessary for evolutionary forces to be significant. By 1900 this mismatch was seen as a major problem. The discovery of radioactivity led to renewed optimism, with Rutherford going so far as to declare in 1904, *"The discovery of the radioactive elements, which in their disintegration liber-ate enormous amounts of energy, thus increases the possible limit of the duration of life on this planet, and allows the time claimed by the geolo-gist and biologist for the process of evolution."* But radioactivity was also soon shown to be insufficient.

In 1929 Houtermans joined forces with Robert Atkinson, a young British astronomer visiting Berlin, who was knowledgeable about the pressure and temperature conditions inside stars. With a little extra help from Geo, they began to examine if quantum tunneling might provide an answer to the dilemma of how energy was formed in stars. The results were suggestive but inconclusive. There was still too little known about nuclei at the time, but the problem would be very much on Geo's mind during the coming decade. He would then play a key role in this subject's development, both by formulating ideas and by prodding others to consider how nuclear physics and stellar dynamics are intertwined.

6 Max's and Geo's Early Careers

I had not felt that I was doing well in astronomy and I did not feel
that I was doing well in physics; and I was just hoping that something would happen that I was doing well and was willing to carry
on with.

Max, on how he felt in 1930

As we have already seen, Geo was well known in the physics world by the end of 1928 thanks to his calculation demonstrating
how nuclei can decay by emitting alpha particles. Max, on the other
hand, had done very little physics research that attracted attention.
Admittedly he was two years younger than Geo, but at age twenty-
two he still hadn't even settled on a research topic, much less written
a thesis. Born could not help him much; overworked and beset with
marital problems, he was on the verge of a nervous breakdown and
about to retreat to a sanatorium. He was still not well after returning
to Göttingen. Luckily for Max, Born's new assistant, the twenty-four-
year-old Walter Heitler, was willing to guide him in writing a thesis.

Together with Fritz London, another young theoretical physicist,
Heitler had just completed a paper containing what we now consider
the first real application of quantum mechanics to chemistry: how two
hydrogen atoms join together to form the simplest of all molecules.
He suggested that Max try to extend their technique to the binding
of two lithium atoms. Lithium resembles hydrogen in its chemical

action, though it is a more complicated element, with three electrons rather than a single one. It wasn't a very exciting problem and, moreover, it was unlikely to lead to any important new insights, but it would at least provide solid training for Max to learn the ins and outs of quantum mechanics. Max set to work and soon found himself bogged down in long, tedious mathematical calculations. As happens with many theses, the results were inconclusive, but the effort was deemed acceptable as the research requirement for a degree. It did not, however, make much of an impression on the outside world.

John Lennard-Jones, a senior physics professor from Bristol, visited Göttingen while Max was completing his lithium calculations. Having funds at his disposal for a one-year research fellowship, Lennard-Jones offered Max a position, hoping the younger man would provide some assistance in the quantum mechanical calculations he himself was pursuing. Quantum mechanics was still very new, but Lennard-Jones realized what an important development it was and thought, probably correctly, that Max was by now more experienced than he in conducting research on the subject. The offer was quickly accepted, and Max set off for Bristol. The move was a culture shock for him—he knew almost no English and had never been out of Germany—but he immersed himself in the local way of life, lived with a British family, soon made new friends, and came to love England. He did travel extensively, but unlike Geo, he did not roam the countryside on a motorcycle in golfing clothes. It seemed, and this would be true in later years as well, that while Max often made an effort to fit in, Geo made one to stand out.

Thinking back on that time in Bristol many years later, Delbrück said, *"I wrote one paper there on the quantum mechanics of inert gases. A very poor paper; I wasn't ready."* He presumably meant that he didn't yet have the tools or the ability to tackle a large-scale problem. In addition, he had an unexpected setback. Having left for England before satisfying all the requirements for a doctoral degree, he returned to Göttingen in December 1929 to take the final oral examination. He thought this was a mere formality, with no serious questions addressed to the candidate and no test of overall knowledge of physics required; he was wrong. The examining committee included an experimental physicist, who was appalled by Max's apparent total ignorance of his subject. Every question he asked elicited in response a calm *"I don't know."* Max failed the exam and was forced to take it again the following year, a good part of which he spent back in Göttingen, studying and making sure the fiasco was not repeated. It was one more example of his not paying much attention to the niceties or formalities of German academic life.

In spite of Max's less-than-sterling performance on the oral examination and his lack of success in research, his senior mentors, principally his own cousin Karl Friedrich Bonhoeffer and Max Born, recommended him for one of the coveted Rockefeller Foundation fellowships that had been designed to support a recipient for a year of independent study. At this stage, Max had only three publications: a small note on Wigner's work, a not very interesting thesis, and what he later called *"a very poor paper."* But it is clear that the older scientists saw something in him; they probably had gleaned from conversations with him a creativity and originality that augured well for the future, if he could be supported until his talents began to bear fruit.

As Max said in an interview decades later, *"I had not felt that I was doing well in astronomy and I did not feel that I was doing well in physics; and I was just hoping that something would happen that I was doing well and was willing to carry on with."* This period was particularly discouraging for Max. The discovery of quantum mechanics had ushered in a dramatic new era in physics; myriad problems that had seemed intractable were now being solved. Novel approaches in such disparate areas as molecular binding, nuclear decay, electrical conduction, magnetism, and electromagnetic radiation were all yielding exciting results. There was an abundance of riches; it was the great revolution in twentieth-century physics. The new generation was being led into mining this new vein by the four new physics geniuses, Pauli,

Dirac, Fermi, and Heisenberg, all of them approaching or just having turned thirty. Max's cohort, a few years younger than the four, was also making a name for itself. Gamow, Lev Landau, Eugene Wigner, Pascual Jordan, Ettore Majorana, J. Robert Oppenheimer, Walter Heitler, Fritz London, Nevill Mott, Victor Weisskopf, Maria Goeppert-Mayer, Felix Bloch, Rudolf Peierls, Hans Bethe—all had published notable papers by this time, and indeed all of them would go on to prominent careers, a half dozen of them winning the Nobel Prize in Physics. But at this point one would have been hard pressed to include Max on the roster.

Then good fortune stepped in: the Rockefeller Foundation granted Max a fellowship. This left him with the decision of how to split the year, the traditional way the award was used. One half was obvious: all young theoretical physicists wanted to spend some time in Copenhagen at Bohr's Institute. Everyone who had been there reported it was a great experience, marked by camaraderie, much fun, and non-stop discussion about quantum theory, led of course by Niels Bohr. As 1930 drew to a close, Max was looking ahead to what he might accomplish in Copenhagen.

While Max's career was showing little progress by this point, Geo's was advancing rapidly. After his initial success with nuclear decay, he attacked the problem of how and why nuclei hold together in the first place, beginning by asking himself why one nucleus is more tightly bound than another. In doing so, he used the analogy of a drop of liquid formed by the cohesion of many small droplets; in this case the droplets would be constituent alpha particles.

With very simple mathematics, Geo proceeded to develop an image of what was happening inside the nucleus. His considerations were like much of his later work: conceptually brilliant, remarkably ahead of what others were thinking, and not entirely correct. His picture of the nucleus was formulated at a time when the neutron, the principal component of a nucleus other than the proton, had not yet been discovered, so the details of his model were necessarily wrong. However, the analogy to a liquid drop was correct and did not disappear.

Developed fully by Bohr and his co-workers in the mid-1930s, it was at the heart of an insight on Christmas Day of 1938 by Otto Frisch and his aunt Lise Meitner. They realized that under certain circumstances, just as a disturbed drop can split, a heavy nucleus absorbing a neutron might break up into two smaller nuclei, a process they named nuclear fission. This was the first step toward the atom bomb.

If Geo had continued his work on nuclear structure, adapting it to new discoveries and reworking his earlier notions, he might have been identified as one of the founders of nuclear structure theory. This is the way science is usually done. An individual has a good idea, stakes his or her claim to it, and pursues its ramifications for years. But that wasn't Geo. As he said, *"I like the pioneering thing,"* so he was always moving on to something else. Consequently, as would happen to him repeatedly in his career, he was given less credit than he would have received had he not continued to look for new and different challenges.

While in Cambridge in 1930, Geo undertook another of those new challenges, writing a nuclear physics text. It seems remarkable that a twenty-six-year-old would have the courage, or should we say chutzpah, to write a book about a subject still so much in flux, but reticence was never a worry of Geo's. Nor was a questionable command of English an obstacle. (Assigned the task of making the text legible for the publisher, Oxford University Press, the book's editor commented wryly that there seemed to be an occasional correct sentence.) However, Geo did have to face a major physics problem, a crucial flaw in the thinking about nuclei in the era before the discovery of the neutron. He stated the prevailing view on page one: *"In accordance with the concepts of modern physics, we assume that all nuclei are built up of elementary particles—protons and electrons."* Aware of this picture's shortcomings, however, he warned readers by using a special skull and crossbones notation at the top of sections in the book that discussed electrons inside the nucleus. When the press suggested replacing the icon with something less ominous, such as a boldface capital letter,

Geo acquiesced, replying to Oxford, *"It has never been my intention to scare the poor readers more than the text will undoubtedly do."*

Once Geo's stay in Cambridge was over, Bohr asked him back to Copenhagen for an additional year, and he readily agreed. Returning there in the fall of 1930, he spent much of his time writing the book, but he also engaged in physics research, most particularly with Max Delbrück, a newcomer to Copenhagen.

7 Copenhagen, 1931

I really longed for him to come back; it was a great vacuum after he
had left because he was such a tremendous vital force.

Max, on Geo in Copenhagen

Upon his arrival in Copenhagen in February 1931 with his
Rockefeller Foundation fellowship, Max was installed in the boardinghouse a few blocks from Bohr's Institute that had become the
favorite place for visiting young physicists. It was so crowded, however, that he had to share sleeping quarters. On his first evening in
the city, Geo, his assigned roommate, organized a trip to the movies
for Bohr and all the postdoctoral fellows, introducing Max immediately to the jolly Copenhagen camaraderie. Perhaps Max even added
a dash to it by wearing the black bowler he had recently acquired in
England, though given Bohr's predilection for Westerns, a cowboy hat
might have been more appropriate. A few years later, on the occasion
of Bohr's fiftieth birthday celebration, one of the young physicists,
Hendrik Casimir, composed a long poem celebrating such occasions.
It begins

> *We went to the flicks, and Bohr came along*
> *And we watched the Black Rider, a man bold and strong,*
> *In a Western picture where guns often bark,*
> *But it's always the hero who first hits the mark.*

At the end of the movie Niels Bohr, deeply moved,
Set out to explain what the plot really proved.

The poem goes on to narrate Bohr's applying to Western shoot-outs
the same type of notions he faced in addressing quantum mechanics,
such as the precision of observations.

Max soon got his own room, but this did not win him much privacy,
since the irrepressible Geo would arrive there late at night and turn
on the lights without any regard for Max's efforts at sleep. Drinking
beer and eating sausages, Geo would begin to discuss physics ideas
that had just occurred to him. Max would sometimes protest, but Geo
was unfazed, and despite occasional protestations, Max was delighted
by his presence. He had never met anybody quite like this Russian.

Geo was an inveterate prankster, an organizer of games and humorous skits. He was also a skilled cartoonist, adorning most of his letters
and the occasional blackboard with comical sketches. He had already
begun a tradition of putting at least one joke in each of his physics
articles, often trying to sneak it by the journal's editors. He first tried
this in the summer of 1930, while preparing a report for *Nature*. He
waited to submit the manuscript until after a weekend climbing expe-
dition with two fellow physicists to the Piz da Daint, a peak in the
Swiss Alps. Sitting on the summit, he penned in the final equation
and added a sentence at the article's conclusion: *"I am grateful to drs.
R. E. Peierls and D. L. Rosenfeld for the opportunity to work here—Piz
da Daint Switzerland July 25."** Lest there be any question, he'd even
had someone take a picture of him on top of the peak signing and
dating the paper.

A year later Geo learned to his dismay that three young German postdoctoral fellows had just pulled off a prank that was more
resounding than anything he had ever done. While in residence at

----

*Geo found the reference to the summit particularly funny, because *pizda* is a Russian slang word for a woman's sexual organs.

Cambridge, they played a joke on Sir Arthur Eddington, perhaps the world's leading astrophysicist, and one of that ancient university's most distinguished professors. In the late 1920s, Eddington increasingly turned away from the kind of work that had made him famous, aiming instead for a synthesis of relativity, quantum mechanics, and gravitation; he called it fundamental theory. The notion was an impressive dream, but it seemed at best without basis and at worst an effort at self-aggrandizement. Eddington did not improve matters by having his attempt based on curious predictions of fundamental constants, the most dramatic being that the strength of the coupling of electric charges to the radiation field had to be exactly the inverse of 136. When better measurements showed that the number was closer to 137, he changed his prediction accordingly, leading some critics to nickname him Sir Arthur Adding-One (by the way, the latest measurement gives us 137.0359976, to within a small error).

The three young Germans wrote a parody of Eddington's work, combining many obviously unrelated physics notions to reach the value of 137: hexagonal crystal lattices, the value of the absolute zero of temperature on the centigrade scale, and the most recent conjectures about relativistic quantum mechanics. The three submitted the nonsensical piece in late 1930 for publication to the preeminent German journal *Naturwissenschaften*. Its unsuspecting editor accepted it immediately, knowing that the three were among the brightest young stars of German theoretical physics and believing that anything they wrote was likely to be worthy of attention. The editor was furious when he discovered that the paper was a hoax. He demanded an apology, and quickly received it. On March 6, 1931, he published a comment about what had taken place:

> *The note by Guido Beck, Hans Bethe and Wolfgang Riezler in the January 9th issue of this journal was not meant to be taken seriously. It was intended to characterize a certain class of papers in theoretical physics of recent years, which are purely speculative and*

based on spurious numerical agreements. In a letter received by the editors from these gentlemen they express regret that the formulation they gave this idea was suited to produce misunderstandings.

Wildly jealous of the prank's success, Geo hatched a plan to top it, and asked Max for help. When another paper with implausible results appeared in *Naturwissenschaften,* this time a sincere one, Geo persuaded several friends to write to the journal's distinguished editor, Arnold Berliner. In their letters, they expressed appreciation for his reprimand, condemned the morals of the young, and then went on to deplore the fact that he had been victimized again. Not rising to the bait, Berliner wrote back:

Of course the paper presents an enormously far-fetched idea, extremely unlikely to be true, and quite unverifiable. However I believe one cannot compare a wild speculation, presented as no more than, let us say numerology, with the thing of Beck, Bethe and Riezler, which you rightly describe as a schoolboy prank. For me, the Editor, these short notes are at times a real curse, but I am afraid I cannot discontinue them.

Geo had been foiled but was not discouraged. He would try again later.

Max and Geo hit it off in Copenhagen, discovering that despite their differences in background, they shared a number of qualities. These included brashness, self-confidence, a humorous outlook on life, and of course a penchant for departing from the accepted path. They were soon working together, co-authoring a paper on nuclear relative probabilities of transitions. (For lovers of long German words, the expression for these quantities is *Übergangswahrscheinlichkeiten.*) But all too soon, summer was upon them: Geo's visa was about to expire, and he had to return to Russia.

Geo's original plan had been to spend a few months riding around

Europe on his motorcycle, attend in October 1931 an international meeting on nuclear physics in Rome, proceed from there to Istanbul, and then return to Odessa to see his father. The first step was getting a visa extension, which he hoped would be granted in Copenhagen's Russian embassy. The ambassador told him this was impossible and that Geo needed first to go back to Russia to obtain a new passport. Off he went, not anticipating any complications. He and Max both thought this would be only a short trip. It wasn't. Geo was denied the desired document with no specific reasons being given.

The Rome conference came and went, with no Geo. Max attended and gave the talk, "Quantum Theory of Nuclear Structure," that Geo had been invited to deliver. People started asking where Geo was. There was a good deal of speculation among the conference attendees, many with strong leftist sympathies, about what was happening in Russia and why Geo had not been granted a passport. Did this portend a new era of political interference in science matters? Was Stalin's consolidation of power resulting in a shift in the Soviet Union's attitude toward the West?

Max felt Geo's absence especially deeply. Years later he remembered what it had been like at the time: *"I really longed for him to come back; it was a great vacuum after he had left because he was such a tremendous vital force."*

Max's term as a Rockefeller Foundation fellow was his only long residence in Copenhagen, but he frequently returned to Bohr's Institute during the years that followed, drawn there by his growing closeness to its leader and by his desire to stay abreast of physics developments. At a very minimum, he attended the yearly weeklong physics meeting that brought back many of the old Institute fellows and visitors for a free-for-all discussion of whatever was topical and exciting in the field. Though this was a completely informal gathering, with no set agenda and no proceedings, the likes of Bohr, Pauli, Heisenberg, and Dirac in the lecture hall's front row were enough to ensure that the exchanges would be stimulating.

These meetings, usually held over Easter vacation, had begun taking place in 1929. At Geo's instigation, it was decided in 1931 that the meetings should include an afternoon or evening of entertainment, primarily devoted to younger physicists performing humorous skits poking fun at their seniors. Aside from some amusement, this was designed to ensure maintenance of the tradition of not being overly respectful of elders. Copenhagen veterans later agreed that the best skit was from 1932. It was a parody of Goethe's *Faust*, written primarily by Max, who had come back to Copenhagen for the meeting; he also acted as master of ceremonies. This skit, in which the Lord was cast as Bohr and Mephistopheles as Pauli, allowed Max to portray with humor two men who had deeply influenced him.

Had he been able to get his passport and leave Russia, Geo would certainly have helped Max with the script. He wasn't entirely absent from the meeting, however, since Max, as an homage to his friend, included in the presentation a large sheet hung in the back with a drawing on it of Geo behind bars saying, *"I cannot go to Blegdamsvej,"* referring to the street on which the Institute was located. Geo later contributed to making the skit known to a wider audience, including this author, by obtaining a manuscript copy from Max, translating it into English with the help of his second wife, Barbara, illustrating it, and publishing it as an appendix to his book *Thirty Years That Shook Physics.*

8 Zurich, 1931

Look, Max, why are you such an interesting man and publish such boring papers?

Wolfgang Pauli to Max in 1931

With the arrival of fall in 1931, it was time for Max to leave Copenhagen and take up the second half of his Rockefeller Foundation fellowship. It seemed natural to use the period to have his research guided by one of the four young physics geniuses a few years ahead of him; each of them had recently become a professor, so there were four locations to choose from. Each man's style was different, and Max now tried to think which one of them would most help his development as a scientist. In Cambridge, Dirac emphasized logical coherence and mathematical elegance; Heisenberg, in Leipzig, was noted for his intuition-guided leaps; and Rome's Fermi seemed to have the ability to discover a trick that led to the essence of a problem. Max felt it unlikely he would be able to follow in the footsteps of any of these masters. Pauli, however, favored gathering all the information and then attacking a problem frontally, an approach that Max thought might suit him very well. Realizing this, Max chose to spend the other six months of his fellowship with Pauli in Zurich.

Pauli was both formidably intelligent and knowledgeable. The quip was that you never had to worry about asking him a stupid question because he regarded all questions as stupid. Furthermore, nobody in the world understood all the ins and outs of quantum mechanics, both the experimental facts and the theoretical calculations, as well as he did. Pauli's first *Handbuch der Physik* review of quantum theory had been the gold standard for work done in the field before 1926; in 1933 he came out with a new *Handbuch* review, "The General Principles of Wave Mechanics." He referred to the two articles as his Old Testament and New Testament. While less satisfied with the second one, he maintained it was *"in any case better than any other presentation of quantum mechanics."* Pauli was certainly not resting on his laurels, nor of course was Dirac, Fermi, or Heisenberg.

Speaking to Pauli could be a challenge, for he was notoriously acerbic, and his barbs and insults were directed at everybody, regardless of rank or importance. Born remembered asking Pauli in 1925 if he wished to collaborate on developing Heisenberg's matrix mechanics and being coldly turned down. Pauli, Born's former Göttingen assistant, said to him, *"Yes, I know you are fond of tedious and complicated formalism. You are only going to spoil Heisenberg's physical ideas by your futile mathematics."*

A considerable degree of self-confidence was necessary not to feel crushed, insulted, or both when you consulted Pauli. Stories of his bon mots, many of them funny and more than one directed at himself, were told and retold in the physics community. It became a kind of badge of honor to be criticized by Pauli, or at least it was if he hadn't said to you, as he once said of somebody's work, *"It isn't even wrong."*

Max was not worried about being wounded by Pauli's barbs. He liked Pauli's directness, absence of hypocrisy or formality, and disregard for politeness. Later in life Max would employ many of the same tactics, anecdotally becoming famous for remarking to other scientists, *"I don't believe a word of what you are saying"* or commenting at the end of a more formal presentation, *"this is the worst seminar I have ever heard."* Occasionally, as with Pauli, people were crushed by such remarks, but they also knew that what both Pauli and Max wanted to hear was a spirited defense by the speaker of his or her position. The two were willing to criticize anyone, not changing their tone for the successful and famous. In their pursuit of scientific truth, neither personal feelings nor status interfered with their finding an answer.

Max and Pauli got along very well in Zurich, forging a lifelong personal bond between them. Ten years later, writing to their common friend Victor (Viki) Weisskopf, Pauli said about Max, *"I still feel a strong friendly relationship with him, as so often happens with two problematic characters."* But he was critical of Max as well. Geo remembered that, while the three of them were sitting in a Zurich bar, Pauli had turned to Max, saying, *"Look, Max, why are you such an interesting man and publish such boring papers?"* Max did not know how to answer. It was quite clear that he thought in novel ways and that he wanted to solve big, important problems, but he hadn't yet found a way to begin doing so. Eventually he did, but it took him several years to respond to the challenge.

The Zurich period did not lead to any concrete results for Max, but it was clear that the depth of his thinking impressed Pauli, as it had Bohr. Both men, who struggled so much to refine the understanding of the meaning behind quantum mechanics, were impressed by Max's efforts to do the same. Having been deeply interested in philosophy and having studied a great deal of it while in Göttingen, Max was particularly fascinated by the originality of Bohr's notion of complementarity. Remembering years later what his thoughts on the subject had first been, he said:

> It also motivated me to look at the writings of Kant on causality and to see how Kant, who was so clever and thoughtful, could have overlooked this possibility. So for the first time, and with a real motivation, I looked at Kant and it was very clear that this situation was just utterly removed from anything that Kant had thought of—so there was no doubt that physicists had been pushed into an epistemological situation that nobody had dreamed of before.

But Max also had to deal now with the practicalities of career advancement. From Zurich, he went back to Bristol for a brief stay, while looking for a position that he hoped would provide him with some stability. He received two attractive offers. The first was a two-year post as Pauli's assistant in Zurich, and the second was a five-year position as a theorist attached to Lise Meitner's experimental physics group at the Kaiser Wilhelm Institute for Chemistry in the Dahlem sector of Berlin. Meitner, a very well established physicist, had been a close friend of Bohr's for over a decade and was well known for both her skill as an experimentalist and her close ties to theoretical physics. She had obviously heard Max praised by both Bohr and Pauli, two physicists she held in very high regard.

From the point of view of professional development, it seemed to Max that it would be preferable to spend a couple of years in Zurich, but he hesitated. It was clear that Pauli thought highly of him, even though he had not produced much work of note, but Max was concerned that he would live up neither to Pauli's expectations nor to the demands that would be put on him in Zurich. In addition, he wanted more freedom to pursue new directions and believed he had a much better chance of doing so in Berlin than in Zurich, where Pauli was such a dominating intellect and personality. Meitner had been reluctant to complicate matters by making a competing offer, but Pauli wrote her in May that Max had not made any commitment to him. He suggested that the decision of where to go should be left to Max. Sensing Max's desire to branch out, he added in his note that if Max chose Berlin, *"he would be forced into greater independence in his thinking and that might be good for him."* Pauli would later realize how correct he had been.

The decision was a difficult one for Max. In Berlin he could live at home with his recently widowed mother and have more flexibility in developing a possible new line of research, since the appointment was for five years. However, Pauli was a great master of theoretical physics. In the end Max chose Berlin, writing to Bohr, *"I have accepted Lise*

Meitner's offer to go to Dahlem as her 'family theorist' largely because of the neighborhood of the very fine Kaiser Wilhelm Institut für Biologie." This is Max's first recorded interest in biology. It was probably stimulated, as we shall see, by conversations in Copenhagen with Bohr, but then, Max also had a way of following his own interests without much concern for what others thought or for career advancement.

9 Max, Bohr, and Biology

It was sufficiently intriguing for me, though, to decide to look more deeply specifically into the relation of atomic physics and biology.

Max, on hearing Bohr discuss biology

When he went to Copenhagen in early 1931, Max was still dealing with his father's death two years earlier, which perhaps explains in part why his subsequent association with Bohr has overtones of a father-son, mentor-disciple connection. In a long interview for German television in 1980, when an already ill Max was near the end of his life, he was asked to reflect back on what Bohr had meant to him. Max struggled to reply, finally describing him as a person *"that one would wish to have as a father, particularly when one had a poor relation with one's own father."*

During his years in Berlin, Max's correspondence with Bohr was often on the subject of biology. Bohr had grown up with a keen sensitivity to the subject because his father was a prominent physiologist, having even been considered for the Nobel Prize in Physiology or Medicine. Moreover, the elder Bohr had a strong interest in questions that ranged beyond purely scientific issues, often discussing such matters with his close friend the Copenhagen philosopher Harald Høffding. In 1911, when Bohr's father died of a heart attack, Høffding's obituary for him read, *"It was his task, in one of the most central areas of organic life, to seek the borderline between life and the forces of inorganic nature.*" As children and, later, as young men, Niels and his brother, Harald, listened to these dialogues and were encouraged to form their own opinions. Perhaps because of his father's musings, the question of *"the borderline between life and the forces of inorganic nature"* became increasingly interesting to Bohr as he grew older.

Until the late 1920s, Bohr's energies had been directed at atomic physics, quantum theory, and the establishment of his Institute, but after 1927 his thoughts frequently took a more philosophical turn. That was the year he formulated the complementarity principle, which he regarded as a bedrock constituent of the Copenhagen interpretation of quantum mechanics. The key for Bohr was the realization that the distinction between waves and particles, which had been so clear before quantum mechanics, was a fallacious one. Apparatus could observe a beam of electrons as particles by making accurate measurements of their positions. Alternatively, interference patterns might reveal the beam acting as waves. The key was that both kinds of measurements could not be made simultaneously; it was either/or. As Bohr phrased it, these were *complementary* ways of studying reality, and it was pointless to say that one was more real or more fundamental than the other. One even had to reexamine what it meant to make a measurement. What were the intrinsic limits of any apparatus, limits imposed not by mechanical precision but by the very laws of nature?

The Heisenberg uncertainty principle, also developed in 1927, showed clearly what the limits were for an interesting set of cases. According to this principle, a particle's position could be determined with infinite accuracy, but the process of doing so precluded obtaining simultaneously any knowledge of that same particle's velocity. The converse was also true. Going beyond the extreme case of perfect knowledge of the one and perfect ignorance of the other, Heisenberg showed how the limits on a simultaneous measurement of both position and velocity were related to a number that lay at the root of quantum mechanics: Planck's constant.

These ideas impressed Bohr so deeply that the complementarity principle and its extension became a central theme in his thinking for the rest of his life. At root, Bohr was a philosopher as well as a physicist, and here he had found an idea that had previously been totally unexplored. He now began to think that complementarity might extend to other fields. The precise limits on simultaneous measurements imposed by the uncertainty principle would not exist, but the more general notion of complementarity could prove useful. For example, in studying economics, was it possible to examine supply and demand simultaneously with essentially infinite accuracy? Perhaps not, for the two might be complementary.

All fields needed to be reexamined; even the age-old argument of determinism versus free will might have to be looked at in light of such criteria. Bohr now began to express these views to wider and newer audiences. In 1929, on the occasion of a celebration honoring Max Planck, Bohr tentatively suggested an extension of complementarity to psychology, asking if there might be limits on the simultaneous examination of brain processes and behavioral phenomena, and hoping *"the special occasion will excuse a physicist for venturing into a foreign field. Above all, my purpose has been to give expression to our enthusiasm for the prospects which have opened up for the whole of science."*

Later that year, at a meeting of Scandinavian scientists, Bohr tackled biology, and thus began to turn to the fundamental question that had already preoccupied his father, *"the borderline between life and the forces of inorganic nature."* Might complementarity provide new insights into the relationship between biology and physics? Bohr was noncommittal, but it was hard to escape the suggestions he was making:

> *Before I conclude, it would be natural at such a joint meeting of natural scientists, to touch upon the question as to what light can be thrown upon the problems regarding living organisms by the latest development of our knowledge of atomic phenomena which I have here described.*

Discussions of this sort were certainly taking place in Copenhagen during Max's 1931 stay. After the successes in quantum theory, these were heady times for the Copenhagen group, and Max, with his own interests in both philosophy and science, found discussions with Bohr stimulating. However, they were also frustrating. Max later reminisced about Bohr's views on complementarity:

He [Bohr] talked about that in biology and in psychology, in moral philosophy, in anthropology, in political science, and so on, in various degrees of vagueness, which I found both fascinating and very disturbing, because it was always so vague. It was vague largely because the situation wasn't clear enough and also in many respects Bohr wasn't sufficiently familiar with the status of the science. So it was intriguing and annoying at the same time. It was sufficiently intriguing for me, though, to decide to look more deeply specifically into the relation of atomic physics and biology.

In order to look *more deeply* into this possible relation between physics and biology, Max decided that he would have to learn a great deal more about biology and, in particular, genetics. This realization lay behind his decision in 1932 to go to Berlin rather than Zurich, where, as Pauli's assistant, he necessarily would have had to devote himself 100 percent to physics.

In August 1932 Max went back to Copenhagen for one of his periodic visits. As he got off the overnight train from Berlin early on the morning of the fifteenth, he was greeted at the station by Bohr's close collaborator Léon Rosenfeld. He had been asked by Bohr to bring Max right away to the Danish Parliament, the Rigsdag, where Bohr would be delivering the opening address to an international congress of light therapists, a five-day meeting devoted to the role of light in "Biology, Biophysics and Therapy." Scheduled for 10:00 a.m., the lecture would be held in the presence of the crown prince, the prime minister, and assorted dignitaries, as well as those attending the congress. Bohr wanted Max there because he was going to be speaking on a subject that he knew would interest him.

Bohr had a way of examining questions from all points of view that some found maddening; Schrödinger once described him as *"completely convinced that any understanding in the usual sense of the word is impossible. Therefore the conversation is almost immediately driven into philosophical questions, and soon you no longer really know whether you take the position he is attacking or whether you must really attack the position that he is defending."*

Max felt Bohr's apparent lack of clarity often stemmed from the fact that he was *"very careful never to say anything that could be definitely called wrong; he was so cautious in his formulations."* But in this lecture, "Light and Life," Max thought Bohr committed himself to a view that could be proved wrong *"and that it was a very good thing that he did, because it challenged me to take it so seriously, to follow it up."*

There is no transcript of the lecture, but Bohr later published a version of what he said that day, in *Nature*. In the text, he states that biology and physics are to be treated autonomously. Though the atoms that make up a living organism are no different from those of any other system, the workings of such an organism cannot be explained simply in terms of atomic phenomena. He wrote:

We should doubtless kill an animal if we tried to carry the investigation of its organs so far that we could describe the role played by single atoms in vital functions. In every experiment on living organisms, there must remain an uncertainty as regards the physical conditions to which they are subjected, and the idea suggests itself that the minimal freedom we must allow the organism in this respect is just large enough to permit it, so to say, to hide its ultimate secrets from us. On this view, the existence of life must be considered as an elementary fact that cannot be explained, but must be taken as a starting point in biology.

It was tantalizing to hear Bohr say that *"the existence of life must be considered as an elementary fact that cannot be explained."* Max asked himself if this was true.

Bohr did not seem to be suggesting, as some thought, that understanding life would require undiscovered principles of physics, but he was suggesting that variations on the ideas proposed by complementarity might be needed to understand the phenomena of life and the accompanying limitations imposed on studying it.

Max had been interested in the issues Bohr raised in the lecture for well over a year, certainly after his arrival in Copenhagen with a Rockefeller fellowship and perhaps even earlier. However, Bohr's sharpening of his own thoughts on the issue had a strong effect on him. In addition, Bohr's feeling so strongly that Max should be present at the congress to hear what he had to say was a clear indication that Bohr thought this would be a good direction for Max to pursue.

Though both Bohr and Pauli had been impressed by Max's intellect, they realized he had not yet found a direction he could call his own, and they were trying to help. Putting myself back in the shoes of a physicist about to turn twenty-six, I can imagine some of Max's thoughts in mid-1932. Though he was still very young, it was becoming increasingly clear that his contemporaries or even younger individuals were succeeding in ways he was not. According to Max, his Zurich office mate, Rudolf Peierls, was *"infinitely more competent,"* as were Lev Landau, Hans Bethe, Max's old Göttingen friend Viki Weisskopf, and numerous others. Max had never been exceptionally proficient in mathematics, so useful for a theoretical physicist, nor was he wildly imaginative like Geo.

What were his skills? How might he succeed? What truly interested him? Could the connection between physics and biology provide the answer? It would be ironic, considering he had started by wanting to be an astronomer, but he clearly was still searching to find his way.

While pondering these questions, Max continued his work in physics during the years of his tenure as "house theorist" to Lise Meitner's

Berlin group. In addition to a review article written together with Meitner, there are two research physics publications from this period. One is an attempt to explain some puzzling data obtained by a graduate student of Meitner's; the proposed solution was reworked twenty years later by Hans Bethe and given the name Delbrück scattering, the only common reference to Max in his career as a physicist.

The other effort, a much more ambitious one, was a very long paper on the relationship between statistical mechanics and quantum mechanics. Unfortunately, that work led nowhere. The fifty-page publication Max produced with a collaborator was read by a few interested scholars but had no effect on the course of physics. As Max later quipped, somewhat sarcastically, it was a *"very learned paper."*

In the meantime, while keeping very much up-to-date on the issues raised in the "Light and Life" lecture, Max also began to engage in serious discussions with biologists and to learn more about their way of looking at these problems. The last sentence of a 1934 letter he wrote to Bohr expresses these concerns: *"This semester I have arranged a private seminar with biologists, biochemists and physicists in which rather able people participate, so that all of us learn a lot."* The *"able people"* he was referring to, about eight or nine individuals, included Kurt Zimmer, a young radiation expert, and Nikolai Timofeev-Ressovsky, a very imaginative geneticist.

To appreciate the efforts Max was making, it is useful to compare his approach to genetics with that of Pascual Jordan, another of the young physics geniuses from the 1920s. Born in 1902, Jordan had done groundbreaking work on quantum mechanics, much of it in Göttingen, while collaborating with Born and Heisenberg. In turning to biology, Jordan was looking to quantum mechanics for ideas that could be extended, but, unlike Max, he was doing so without making the effort to actually learn the details of the new field. His biology arguments amounted to a kind of vitalism, the notion that the origin of life is due to a vital force rather than the ordinary forces of physics and chemistry. He argued that random atomic reactions were amplified coherently in animate matter. According to Jordan, conscious life emerged as a consequence, but the possibility of observing directly the amplification was probably undetectable because of limitations on measurements imposed by quantum theory principles.

When, late in 1934, Jordan addressed a Berlin meeting of the Society of Empirical Philosophy, he claimed to represent Bohr's views on the subject of quantum mechanics and biology as well as his own. Max objected, writing Bohr that he had prepared a statement about what he believed Bohr's views actually were and had given it to a prominent Berlin biologist, who had *"complained bitterly at the meeting about the confusion in the biological literature caused by your and Jordan's views."*

Unlike Jordan, Max favored what we call a bottom-up rather than a top-down approach: first find out what the problems are in the new field, learn the nitty-gritty of the experimental data, and then see if new ideas from the field you are familiar with might be useful. Don't start with preconceived notions of what is right. This turned out to be a far wiser approach than Jordan's.

10 Max, Berlin, and Biology

Gene mutation, as an elementary process in the sense of quantum theory, in particular as a specific change in a complicated assemblage of atoms, can be considered secure.

Max, writing in 1935

Max was lucky to find in Timofeev-Ressovsky the perfect collaborator for his initial forays into genetics. Nicknamed in Russia the Wild Boar, Timofeev-Ressovsky had arrived in Berlin in 1925 through an exchange between Moscow and Berlin, and by the early 1930s he had established a reputation as a leading researcher. Genetics had only recently begun to be linked with physics through the study of mutations induced by exposure to X rays. Though Timofeev-Ressovsky was an expert in these new analyses, he was relatively ignorant of the insights that quantum techniques might provide and was therefore eager to collaborate with a physicist who was knowledgeable about such matters.

The origins of genetics are usually traced back to the work of Gregor Mendel, an Austro-Hungarian Empire monk whose patient observations of peas grown in his abbey's garden allowed him to deduce the rules regulating how traits are passed from generation to generation. However, Mendel's conclusions, published in 1865 in an obscure journal, were ignored for thirty-five years, so the field's origin is usually taken to be 1900, the date of the rediscovery of Mendel's work.

A major advance in genetics research took place in 1907, when Columbia University professor Thomas Hunt Morgan began experimenting with *Drosophila*, commonly known as fruit flies. *Drosophila melanogaster* are tiny yellowish-brown insects that live happily at room temperature, feast on a steady diet of rotten bananas, reproduce every few days, and are large enough for their many mutations to be visible to the naked eye. By the early 1930s they were the preferred instruments for genetic studies. Morgan's first indicator of a mutation came in 1910, when he discovered a fly with a white eye instead of the standard red. Activities in his laboratory, soon known as the Fly Room, took off rapidly after that.

Since life begins with the linking of chromosomes, the next step was to find the location along them of the genes that determined particular mutations: in other words, to identify the ones responsible for a varied form of life. In 1911 Alfred Sturtevant, a brilliant nineteen-year-old mathematically inclined student of Morgan's, managed to deduce the ordering along the chromosome of the known fruit fly mutations. However, this still provided few clues to solving the riddle of what a gene actually is. The simple answer, a tautology, was that the gene is the unit of heredity. But was it possible to understand its structure? What was it in a gene that led to mutations?

Hermann Muller, another veteran of Morgan's Fly Room, provided a partial answer in the 1920s by discovering that irradiation with X rays significantly boosted the mutation rate in fruit flies; continuing to increase the dosage eventually killed them. This work garnered a great deal of attention, both for its intrinsic scientific worth and for the warning it gave medical practitioners and others who were indiscriminately using the newly discovered rays. Unfortunately, in the United States of the late 1920s and early 1930s, Muller was experiencing numerous difficulties that had little to do with science. Threatened with dismissal from his university position for his outspoken leftist views, he decided it was wise to leave the United States. Having met Timofeev-Ressovsky on a visit to Moscow in the early 1920s and introduced him to the findings from the Fly Room, Muller knew he would be warmly welcomed in Berlin, so he moved there in 1932.

Muller and Timofeev-Ressovsky quickly set to work correlating X-ray doses and wavelengths with mutation probabilities in both fruit flies and bacteria. In the process, they found some strange results. They observed that organisms subjected to conventional chemical poisons survived until a threshold was reached, at which point they all died, but the lethal effects of X rays showed no sharp transition. The key seemed to be how the radiation was directed and focused, not its total amount. Trying to interpret these results, the two experimenters found themselves using a language that was new to biology. Mutations occurred only after a certain number of "hits" on a "target." But how many hits were necessary and how large was the target? This was beginning to sound like the physics experiments that Rutherford and his collaborators had done with beams and particle counters. Because of the resemblance, Muller urged Timofeev-Ressovsky to look into the possibility of collaborating with physicists, but who among them would be knowledgeable and flexible? They needed technical and conceptual help, not a lecture on quantum mechanics. Hearing of Bohr's interest in biology, Muller made a trip to Copenhagen in the spring of 1933, but he was disappointed by the encounter, writing a friend that he had been glad to *meet the physicist Bohr there, but then I found that his ideas in biology were hopelessly vitalistic.*

Muller's stop in Berlin turned out to be short, his departure almost coinciding with Max's return to Berlin. Alarmed by Hitler's ascent to power in January 1933, Muller decided to relocate once more, accepting an offer to transfer his work to the Soviet Union. That stay was also not a very happy one. Russian genetics research, under the direction of the ill-informed and malevolent Trofim Lysenko, was descending into chaos, following party lines rather than scientific truth. Persecutions began, and so Muller was again on the move, finally returning to the United States at the beginning of World War II.

Meanwhile, back in Berlin, Max contacted Timofeev-Ressovsky because he wanted to organize a series of informal gatherings where scientists with varying backgrounds would focus on biological problems. Finding that he and Timofeev-Ressovsky had a common interest, he began the meetings. Some were held in the old family residence Max was living in with his mother; others, at Timofeev-Ressovsky's house or laboratory. One participant remembers that *"we talked, usually for ten hours or more without any break, taking some food during the session."* Neither Timofeev-Ressovsky nor Max subscribed to the view that the great French biologist François Jacob says most geneticists held at that time, that the gene is a *"three dimensional structure of forbidding complexity, offering no hold for experiment."* But they did realize that progress would require a new kind of approach, one involving expertise from biology, chemistry, and physics. They would need to form a team with each member understanding fully what the other was doing and how he had reached his conclusions.

Timofeev-Ressovsky later remembered his impressions of Max:

> *. . . somewhat insolent, but this is excusable. We also treated him insolently, and he acquired our manners very quickly. . . . I told him about Koltsov's general concept of molecular biology of genes and chromosomes, and our attempts to prove the monomolecularity of the gene. That is to say, this is a joint physico-chemical elementary structure, and not a piece of butter!*

Max's directness, sometimes perceived as arrogance, was nothing new. He was known to act this way even with Bohr, but this behavior clearly did not bother Timofeev-Ressovsky. The two of them were personally fond of each other and soon began to work together; Karl Zimmer, a young German who had received his doctorate in Berlin the year before with a thesis on photobiology, soon joined them. Their three-way collaboration led in 1935 to an influential paper. Zimmer was the radiation expert, Timofeev-Ressovsky the geneticist, and Max was responsible for interpreting the results within a theoretical framework that would be consistent with quantum mechanical principles.

It is still very unusual to see a science paper written by three authors from three different disciplines. At the time, it was essentially unheard of, but this was exactly the synergy Max had hoped for when he opted to go to Berlin instead of Zurich. Entitled "On the Nature of Gene Mutation and Gene Structure" and sometimes referred to as the Three-Man-Labor (*DreiMännerArbeit*), the paper has, as we shall see, become famous in genetics history, perhaps less for its contents than for the influence it had on Schrödinger and, through him, on a younger generation of would-be scientists who read a small book he wrote entitled *What Is Life?*

Part one of the fifty-page opus was by Timofeev-Ressovsky, part two by Zimmer, and part three, suggestively subtitled "Atomic Physics Model of Gene Mutation," by Max, with all three joining to write the conclusion. The aim, set down clearly in the article's beginning, was to

> arrive at a theory of the mutation process and of gene structure that will be experimentally well-founded and that will have experimentally verifiable implications. Of course we are far from regarding our presentation as conclusive; rather we see its value lying in the extending of earlier approaches through the use of the concepts of physics.

In short, the paper was saying that *"concepts of physics"* would help explain gene structure. Its conclusion was, as Delbrück wrote, that *"the view of gene mutation, as an elementary process in the sense of quantum theory, in particular as a specific change in a complicated assemblage of atoms, can be considered secure,"* where the word *secure* carries the meaning of "convincingly demonstrated." In other words, a mutation, the means by which, for example, a red-eyed fruit fly produces offspring with white eyes, was to be regarded as caused by either a rearrangement of the atoms in the molecule or a direct change in the structure of the constituent atoms. In either case, an X ray striking the molecule provided the missing energy necessary for the shift to be made. These processes, regulated by quantum theory rules, were well understood. The details of what transpired might be very complicated, but no new concepts needed to be invoked.

The paper appeared in a journal that very few scientists read, but the three authors were given a set of reprints. They mailed these to people they thought might be interested in knowing about their conclusions. Max, of course, sent one to Bohr, with a covering letter saying with some regret that there seemed to be no need for the kind of notions Bohr was so fond of. Complementarity did not apply, because the transition from one atomic or molecular configuration to the other was straightforward. He wrote Bohr:

> *The paper contains no complementarity arguments at all. On the contrary it turns out that one can formulate a unified atomic-physical theory of mutation and molecular stability. This is due to the fact that we do not need to know anything at all about the precise way in which genes act in the developmental process.*

The paper is famous because the three authors, proceeding logically, had succeeded in demystifying the nature of genetic mutations. Still, they had not put to rest the question of whether new laws of physics would be needed to explain what made an organism *be alive*, the question that had concerned Bohr and made him consider if some extension of complementarity was necessary. They had only provided a model for how a living organism could change from one form to another—that is, mutate. But that was already a great deal.

11 Geo Escapes from Russia

Gamow has kept away from politics and social activity. In his behavior, he is relatively undisciplined and is a typical representative of the literary-artistic bohemia.

Soviet police report on Geo

In 1932, as Max was settling back in Berlin and starting his discussions with Timofeev-Ressovsky and Zimmer, Geo was discovering how much Russia had changed while he was away. Matters had been very different in the summer of 1929, on his first visit home after a year in the West. He had then been received triumphantly, his theory of nuclear decay a symbol of how Russians could compete successfully in the science world. Newspapers had announced, *"A son of the working class has explained the tiniest piece of machinery in the world: the nucleus of an atom,"* and the Communist Party's official organ, *Pravda,* had praised his achievements on page one. Two years later, however, times had changed. Stalinism was now beginning to impose itself as a repressive force even in the abstract domain of physics. A distinction was now being made between capitalist and proletarian science, with not so subtle hints that Russian scientists should learn the secrets of the former while not revealing any of the latter's to colleagues in other countries. In addition, suspicion of anyone expressing a desire to go abroad was becoming rampant.

When Geo gave a popular lecture on quantum mechanics to the

House of Scientists, he was stopped when he began discussing the Heisenberg uncertainty relations. The audience was dismissed, and he was ordered to never mention them again. They supposedly did not conform to the tenets of dialectical materialism.

Meanwhile, to make ends meet in the impoverished Soviet Union, Gamow had patched together five separate appointments, which entailed his standing in line at five separate offices to collect his salaries. It also meant there was a new excuse for not allowing him to leave: he was too valuable, too needed by Russian proletarian science. On the other hand, he was not above suspicion; party officials, requesting information on him from his employers in 1932, were told, *"Gamow has kept away from politics and social activity. In his behavior, he is relatively undisciplined and is a typical representative of the literary-artistic bohemia."* Though this was not the sort of condemnation that would lead to immediate trouble in Russia in 1932, it did not bode well for his future.

Realizing that he was not likely to be granted a new passport, Geo began to think of ways he might escape without one. The decision was heart-wrenching because he loved Russia. Once, when asked why his autobiography essentially ends with his departure from his native country, Geo responded, *"Nothing of interest happened to me after that."* However, despite how much Russia meant to him, his fear of staying was becoming greater. He later wrote about why he had left:

> *I had always felt that I did not want to desert my native country, and that as long as I was permitted to travel beyond the Soviet borders and keep in contact with world science, I would always come back home. I could not possibly accept the theory of the alleged hostility between "proletarian" and "capitalistic" science; it just did not make any sense to me. Also the increasing pressure of the dialectical-materialist philosophy was too strong, and I did not want to be sent to a concentration camp in Siberia because of my*

views about the world-ether, the quantum mechanical uncertainty principle or chromosomic heredity.

The prospects for leaving were complicated in late 1931 by Geo's meeting and soon marrying the beautiful Lyubov Vokhminzeva, a former physics student working in an optics laboratory. Nicknamed Rho, she was also a ballerina—their son, Igor, told me of her devotion to the art. With fuel scarce and a minimum of warmth necessary, attending ballet school in winter meant each student traveled there every day with costume and shoes under one arm and a log to heat the studio's woodstove under the other.

Despite her own attachment to Russia, Rho was as eager to flee as her new husband. The obvious escape route was the nearby Finnish border, but all its crossings were heavily guarded. Venturing into Norway north of the Arctic Circle or Persia in the Caucasus turned out to be not much more promising. Having discovered difficulties on all the land tracks, Geo and Rho turned to the possibility of finding a way out over water.

The Crimean peninsula, jutting out into the Black Sea, seemed to afford the most promising choice. Only 170 miles separate its southernmost tip, near Yalta, from the Turkish coast, an easy journey in a fast boat. Growing up in Odessa, Geo was very familiar with the constant back-and-forth travels of smugglers crossing those waters with European luxury goods, which then made their way to Moscow or Petrograd, but that trafficking had been eliminated. Even fishermen were now subject to careful scrutiny, their boats repeatedly examined, and permission to go out at night was rarely, if ever, granted; this avenue seemed closed.

Geo and Rho then heard that a Moscow factory was producing a collapsible two-person boat resembling the kayaks the Aleuts used in the frigid Arctic waters. With a rubber hull held in place by sticks, the boat was intended for sporting trips and could be taken apart. It was also light enough to be carried in a pair of knapsacks. But was such a trip possible? It might take five or six days, a seemingly impossible journey under the best of circumstances, and theirs was a far cry from that, since all planning and supplying had to be carried out in secret. They could not even book a hotel room on the coast: such facilities were reserved for foreigners and party functionaries. However, there was a House of Scientists on the Crimean shore where Geo could have a room for vacation purposes, and Rho, who did not have permission to stay in it, was nevertheless allowed to share meals with Geo and use the facilities; she needed to be lodged in a nearby inland village.

Deciding that it was worth the risk, Geo and Rho set their exit plan in motion, purchased the boat, and took it to the Crimea, supposedly for recreational purposes. Having assembled the boat on the shore, they embarked on a series of short sorties to study paddling techniques, storing the boat with a fisherman so as to be ready to go when their odyssey began. Food was also a problem. Shortages existed all over Russia, and Geo and Rho were limited as to what to bring, certainly nothing too bulky. The solution they found was to gradually purchase eggs, which were rationed, and hard-boil them. These, coupled with some bricks of cooking chocolate, two bottles of brandy, and drinking water, constituted their food supplies. A final discussion centered on Rho's decision to bring her toothbrush.

Another obstacle was what to do should they succeed in reaching Turkey. The last thing they wanted was to be identified as Russians and repatriated, an outcome that would doubtless lead to both of them being imprisoned. Geo's solution, even more improbable than his escape plan, was to carry no documents except his by-now-expired Danish motorcycle license. Landing on the shores of Turkey, they would try to pass themselves off as Danes. With luck, they would be taken to the Danish embassy, and from there, Geo would call Niels Bohr. After that, all would supposedly be well.

Geo and Rho set out early on a clear 1932 summer morning, intending to steer south during the day by using a compass and navigate at night by the North Star. They would take turns paddling, switching every half hour. The first day in their kayak-like boat went well, but a storm came up during the night, and dawn found them struggling to avoid being swamped by the big waves. Geo valiantly attempted to keep the boat from capsizing, and Rho bailed out the water seeping in as the waves rolled over them. By nightfall the storm had subsided, but the two were completely exhausted. Beginning to hallucinate, Geo remembered restraining himself from getting out of the boat and walking away on the moonlit glassy surface of the sea, while Rho conversed with a fantasy ship captain. Late that night they dimly saw a sandy shore, struggled toward it with their last strength, and collapsed on the beach. Awakening, they saw several Tatar fishermen staring at them, but they were unable even to stand up. They had landed only a few miles from where they had departed forty-eight hours earlier.

The fishermen took them to a hospital. After a brief recuperation, they returned to the House of Scientists, explaining that they had been blown off their intended course by the sudden storm. Nobody suspected they had tried anything as foolish as paddling to Turkey.

In his autobiography Geo describes their adventure as *"very childish, and in fact we behaved very irresponsibly throughout the whole affair."* Despite their optimism, the chances of crossing the Black Sea in a kayak seem slim under the best of circumstances. Still, desperate people have always been willing to take great chances to escape. Fortunately, unlike so many others, this story had a happy ending. A little over a year later, Geo and Rho managed to escape in the most banal way, and Bohr played a role in helping them exit from Russia.

The small, elite Solvay Conference, held as always in Brussels, was set for October 1933, with the announced topic "Structure and Properties of Atomic Nuclei." The world's leading physicists in the subject were expected to attend, and given Geo's prominence in the field, it was natural that he would be asked to participate. Bohr astutely realized that a simple invitation would probably not be enough for Geo to be given a passport, so he asked Paul Langevin, the conference's convener, to place a direct request with the Kremlin that Geo be sent as Russia's representative. This scheme worked. Geo obtained the desired passport and, miraculously, was even allowed to have Rho accompany him.

The result was that Geo and Rho went off to Brussels in October 1933, planning not to return to Russia if at all possible. When he heard of their intended defection, Bohr became concerned that it might compromise Langevin's relations with the Soviet regime; he urged Geo to at least consider returning to Russia. Pouring his heart out to Madame Curie at a dinner held in Paris immediately after the Solvay Conference, Geo asked her what he should do. Curie as a young woman had fled to France from Russian-dominated Poland and was particularly sympathetic to the sufferings inflicted on young Russians. She told Geo not to worry; she would speak to Langevin, her friend and colleague.

The next day she found Geo waiting to meet with her in the reading room of the institute she directed. She informed him that he need not be concerned about Langevin's relations with the Soviet Union and that, furthermore, she was willing to provide Geo with a stipend for two months while he looked for a position in the West. During those two months in Paris, Geo was contacted by Rutherford from Cambridge and Bohr from Copenhagen, each of whom provided extra months of support, enough to tide him over until he could take up a summer school lectureship he had been offered at the University of Michigan.

Borrowing money from Bohr and Rutherford for the transatlantic passage, Geo and Rho boarded a ship bound for New York City in the early summer of 1934. Upon their arrival in the United States, Geo was contacted by the president of George Washington University, who offered him a one-year visiting professorship at his institution and a promise to make the position permanent after that. Geo accepted happily. He and Rho were finally safe.

12 The Russia Geo Left Behind

It brings back to memory the idyllic and enthusiastic sessions at your house and at our house where we delighted in our first adven-tures at bringing genetics and physics together.

Max to Nikolai Timofeev-Ressovsky, on their 1935 collaboration

Reflecting on what happened in Russia after he left, Geo was sure that his sense of humor and his inability to toe the line politically would have proved a deadly combination. Lev Landau, his dear friend from his Leningrad and Copenhagen days, almost died in a Soviet prison during the late 1930s, despite being a firm believer in communism, and Geo was sure he would have fared no better. Arrested in April 1938, at the height of the Stalin purges, Landau was saved only by the extraordinarily courageous intervention of Pyotr Kapitza, a great Russian physicist, who was himself trapped in Russia against his will.

Kapitza had arrived in Cambridge, England, in 1921, soon becoming one of the chief experimenters there and a particular favorite of Rutherford's. He and Geo must have formed quite a pair in late-1920s Cambridge, each wildly colorful by local standards: crazy Russians on motorcycles. By 1934 Kapitza was a Royal Society professor and the director of a new laboratory dedicated to exploring properties of matter at very low temperatures. That summer, on his yearly visit back to the Soviet Union, he was detained, denied a passport, and denied further travel abroad. The Soviet authorities, afraid he would not come back if they let him out for even a short time, were not going to make the same mistake they had made with Geo.

When Rutherford protested that Kapitza should be allowed to return to the West for the good of science, the Soviet embassy reply was chillingly clear in its bureaucratic language:

> As a result of the extraordinary development of the national economy of the U.S.S.R., the number of scientific workers available does not suffice and in these circumstances the Soviet Government has found it necessary to utilize for scientific activity within the country the services of Soviet scientists who have hitherto been working abroad. Kapitza belongs to this category.

The rumor circulating in Russian scientific circles was that when Vyacheslav Molotov, the chairman of the Council of People's Commissars, was told, *"Don't you know that a bird in a cage won't sing?"* he replied, *"This bird will sing."* To ensure a gracious melody coming forth, the Soviets built Kapitza a brand-new laboratory at a beautiful location in the capital. He was also allowed to make a set of strategic hires aimed at developing a world-class research center. Lev Landau, not yet thirty but already recognized as a great theoretical physicist, was one of the key recruits. Even more than Geo, Landau was unfailingly blunt, respecting intellect not status. This attitude, not a problem at Bohr's Institute, was recklessly dangerous in 1938 Russia. Landau knew this but was unable or unwilling to change.

When Landau was arrested, Kapitza wrote directly to Stalin, despite being himself under suspicion. Adopting a tone that he hoped would assuage the dictator's concerns, he acknowledged that Landau

> is a trouble-maker, he likes to look for shortcomings in other people, and when he finds them, particularly in high-ranking elders, such

as Academicians, he starts mocking them most disrespectfully. This has made him many enemies.

Kapitza concluded the letter by saying,

I nevertheless cannot readily believe that he would be capable of anything dishonorable. Landau is young, and he has still much to do in science. Only another scientist could write about all this, and this is why I am writing to you.

He never received an answer.

Time went by. Hearing reports that Landau was near death, Kapitza now wrote to Molotov. While acknowledging that *"I am interfering in what is not my business since this matter falls within the province of the Commissariat for Internal Affairs,"* he went on to say that Landau was *"in poor health, and if he should perish to no purpose, it will be a most shameful thing for us, the Soviet Union."* As one friend of Landau's wrote, *"Everyone knew at that time that to do such a thing required more courage than going into a tiger's cage."*

Still, it worked. Lavrenty Beria, the new head of the secret police (NKVD), released Landau under Kapitza's personal guarantee.

Let out of prison after a year, Landau had learned to be guarded in what he said and how he said it, except in strictly scientific matters. He went on to become the founder of a great school of theoretical physics in Moscow, one that survived all the turmoil of the cold war, continuing to shine until well into the 1980s. Seminars were sometimes held in people's apartments; discussions were pursued in parks and on streets as well as in classrooms. Conditions were never easy, but the flame of scientific inquiry burned brightly.

The school even survived the loss of its intellectual leader. Still at the height of his powers, Landau was injured in a 1962 automobile crash, lay in a coma for two months, and never fully recovered. He died in 1968, a few months before Geo did. Though they never saw each other after Geo left Russia, one not being allowed to enter the country and the other to leave it, Geo always regarded "Dau" as his dearest friend, his comrade from their old, carefree days.

Biology, and genetics in particular, fared far worse than physics under Stalin, with the field becoming politicized in a way that physics did not. Russian genetics investigations were very promising until the mid-1930s, largely because of the efforts of Nikolai Vavilov, the leader of modern biology research in the country at the time. However, Trofim Lysenko, eleven years younger than Vavilov and from a peasant rather than a bourgeois family, was maneuvering to replace him as overseer of Russian genetics, claiming that he, rather than Vavilov, represented the true aspirations of the Russian people. Denying the validity of Mendel's findings, and holding instead the Lamarckian view that acquired characteristics could be transmitted, Lysenko propagandized the virtues of proletarian science, exactly the kind of mixing of science and politics that Geo had feared would come to dominate in Russia. Promising to boost Soviet agricultural production through his findings, Lysenko gained direct access to Stalin, and by 1938 he was in control of the branch of science Vavilov had tried to organize. Vavilov sensed that his own imprisonment was imminent. This occurred in 1940; he died in prison three years later.

Timofeev-Ressovsky, Max's old Berlin friend and colleague, also suffered because of Lysenko. Anticipating the impending shift in Russian biology, Vavilov had already sent him in 1937 a secret warning that any return to Russia would likely lead to his arrest, but Timofeev-Ressovsky did not heed it. He remained in Berlin through World War II and refused to flee to the West as the Russians approached the city at the war's end. Despite his twenty years in Germany, he continued to regard himself as a Russian, and persisted in believing he would be treated fairly by his compatriots. The Germans executed Timofeev-Ressovsky's son for anti-Nazi activities, and in 1945 the Soviets executed his brother for anti-Soviet activities. Timofeev-Ressovsky survived, but was deported to a gulag in Kazakhstan. Near death after two years there, he was released because of international pressure and was transferred to a site in the Ural Mountains, where he began work again. Though almost blind because of his sufferings in the gulag, he now focused his attention on the biological effects of radiation.

Timofeev-Ressovsky and Max did meet one more time after the 1930s. When Max was notified in 1969 that he had been awarded the Nobel Prize in Physiology or Medicine, his first reaction was to say to his wife, Manny, *"Good, that will make it easier to go and see Timofeev."* Still, despite his good intentions and the greater visibility the Nobel afforded, Max could do little for his old friend. After visiting him in Moscow, he wrote Timofeev-Ressovsky a letter in which he reminisced about their old days together:

> *It brings back to memory the idyllic and enthusiastic sessions at your house and at our house where we delighted in our first adventures at bringing genetics and physics together. Like the young Ladies and Gentlemen of Boccaccio's Decameron, we were brought together by withdrawing from the terrors of a great plague to jointly consider some of the riddles of life.*

The *"great plague"* of Boccaccio's tales was the Black Death, carried into Florence in 1348 by rats and fleas, whereas the one Max refers to was brought upon Berlin by the Nazis. It must have been a great sadness for Max to see that Timofeev-Ressovsky, who had suffered so long under Hitler, was now languishing under the rule of Soviet dictators.

13   Geo Comes to America

At the crack of dawn—I mean at nine-thirty in the morning—Geo
Gamow would call me almost every day with an idea and that idea
was simply wrong. Almost always!

Edward Teller, on collaborating with Geo in the late 1930s

Safely out of Russia, the Gamows were thriving in their new
environment. Kapitza's being kept in the Soviet Union had reinforced
their decision not to return there even for a visit, lest they be trapped
again. On November 4, 1935, Rho gave birth to a son, Rustem Igor.
The name Igor, derived from an early Russian saga, was a reminder
for Geo and Rho of the homeland they'd left behind, but they viewed
present Russia with little nostalgia. Nor did they try to surround themselves with other émigrés. They were enjoying both the comparative
ease of living in the United States and its acceptance of foreigners.

When recruited by George Washington University, in Washington,
D.C., Geo had asked the administration to support and partially fund a
yearly small, informal conference in theoretical physics that would be
organized around a topic deemed interesting at the time. Bohr's meetings
in Copenhagen, the ones both he and Max had found so stimulating,
were the models for what Geo hoped to achieve. The request was quickly
granted, and Geo began making plans almost at once. The First George
Washington University Theoretical Physics Conference was held in late
April 1935, a time of year when Washington's cherry blossoms are in

bloom and the city is at its loveliest. The gathering, devoted to nuclear physics, had a relatively local character, bringing together physicists from the East Coast, many of them European refugees like Geo. Later, as word spread about how interesting the discussions were, scientists came from farther away. A formidable group, including half a dozen future Nobel Prize winners, attended the second meeting, held at the same time of year as the first one. Its announced topic was molecular studies, chosen in part to highlight the expertise of a new arrival at the university.

Upon joining the faculty, Geo had asked that an additional position in theoretical physics be created within a year. Since he was an expert in nuclear physics, he had a preference for someone with interests oriented toward atomic and molecular physics. He wanted somebody who fit this bill but was also broadly knowledgeable and easy to communicate with. He already had a candidate in mind: a twenty-seven-year-old Hungarian named Edward Teller, who had been a student of Heisenberg's in Leipzig, an assistant of Born's in Göttingen, and a veteran of two stays in Copenhagen, during one of which he and Geo had become friends. This young man had a temporary position in England at the time and was looking for something more permanent. George Washington University agreed that he was a good choice; the offer of a professorship was tendered and quickly accepted.

Teller arrived in August 1935 with Mici, his new wife, and a friendship between the couples developed rapidly. Their dinners together were odd affairs from a linguistic point of view. Geo and Edward spoke together in German, the language they had used when they met in Copenhagen and the working language they employed with each other until 1939. The day the Nazis invaded Poland, Teller switched to English and would never use German again. In those early days in America, however, while the two continued to communicate in German, Geo and Rho would converse in Russian, while Edward and Mici did so in Hungarian, their native tongue. If the two couples wanted to have a four-way conversation, though, it had to be in English, their only common language, and the one they were all struggling to learn. I can only imagine the confusion this caused in restaurants as others dining around them tried to guess what they were speaking, though perhaps Russian, Hungarian, and German all sounded alike to most Americans at the time: foreign.

Geo and Teller settled quickly into an enjoyable working relationship. Teller, a night owl, remembered what it was like:

At the crack of dawn—I mean at nine-thirty in the morning—Geo Gamow would call me almost every day with an idea and that idea was simply wrong. Almost always! Geo Gamow had the wonderful property that he did not mind being wrong. He did not do it for the prestige. He did it for fun. He did it for love. And when his idea was not wrong it was not only right, it was right and new.

And the two of them did have fun together.

Their first joint physics publication was a significant paper about nuclear decay. Geo had made a name for himself in 1928 by developing an explanation for how this could occur through alpha particle emission. Now he and Teller turned their attention to those nuclei that decay primarily by emitting an electron. The processes they were interested in had posed a significant problem in interpretation for many years because they apparently violated the physics bedrock principle of energy conservation. Two opposing explanations had been offered. While Bohr suggested that energy was not conserved within the confines of the nucleus, Pauli, rejecting this notion, proposed that an undetected particle carried off the missing energy. After Fermi's 1934 seminal work demonstrated a mechanism for Pauli's scheme, the mysterious particle's existence had been accepted. It had also acquired a name, chosen by Fermi: neutrino.

Gamow and Teller's work was one of the important topics at the third George Washington University conference, held in 1937. The scientists met in a less welcoming season, mid-February rather than April, but the two new professors at the university were trying to accommodate a very special physicist: Bohr, making a round-the-world trip, would be on the East Coast at that time. He was eager to see how the "Copenhagen spirit" had been transplanted to American soil, to reacquaint himself with old friends, and to meet American physicists he didn't know.

Those attending the conference, whose theme was nuclear structure, issued a special report at its conclusion, saying that the meeting had been a great success in stimulating ideas and suggesting avenues for further research. Moreover, the participants felt it was better than any other meeting they'd gone to.

Still, Geo was restless and beginning to feel that there were too many people studying nuclear reactions. Maybe it was time to leave the subject and do the *"pioneering thing"* again. Could the new discoveries in nuclear physics supply the needed answer to what powered the sun? Was the time now ripe to begin looking again at this problem, considered by him and his friend Houtermans almost a decade earlier?

In the early 1920s, Arthur Eddington had conjectured that Einstein's famous formula relating mass to energy might provide an answer to how the sun managed to produce so much energy. He argued that if four hydrogen atoms could be converted into helium in a stellar interior, and if the mass difference between the two configurations were to reappear as energy, all one would need to fuel the sun would be sufficient hydrogen. However, there were two huge problems the suggestion had to confront. The first was total ignorance of how the conversion could take place, and the second was the belief that the sun's composition was similar to the earth's (that is, containing little or no hydrogen).

In the late 1920s, the astronomers Cecilia Payne and Bengt Strömgren had shown independently that, contrary to previous assumptions, hydrogen and helium were the sun's main constituents. And in 1929 Fritz Houtermans and Robert Atkinson had demonstrated that Geo's tunneling mechanism could, in principle, provide the means for conversion in a stellar core. Too little had then been known about both atomic nuclei and stellar interiors to go any further, but in 1938 Geo, who had helped Houtermans and Atkinson with their original calculations, thought the time was ripe for real progress.

In the beginning of the year, Geo and Teller picked "Problems of Stellar Energy Sources" as the topic for the Fourth George Washington University Theoretical Physics Conference. They decided they would invite both nuclear physicists and astrophysicists to attend, to see if they could create a meeting of minds. This inspired decision, similar in many ways to Max's quest for finding common ground between physicists and biologists, would have a profound and lasting effect on both physics and astronomy.

14   The Sun's Mysteries Revealed

Bethe proved himself the champion at Gamow's game.

Edward Teller, on Bethe's solution of the solar energy problem

We have already seen examples of Max's and Geo's penchant for the *"pioneering thing,"* but another common trait related to that in a number of ways was their continual effort to reach out to scientists in other fields. This came about partially as a result of natural curiosity and partially as a way to help with the *"pioneering thing."* What may be less clear, but will become increasingly obvious as we see their careers develop, is what social animals both Max and Geo were. They viewed science as fun, as play, as a game to be shared with others. There were rules that had to be obeyed, of course, but they believed one must never lose sight of the joy, the privilege, of being able to pursue this strange calling. And part of the fun for them was to bring together all sorts of scientists and see what they came up with, what connections could be made. This side of Max's character had not yet emerged the way it had in Geo, but it would become increasingly evident as Max rose to prominence and made science research a feast of sorts for himself and those around him. In this respect he and Geo were more similar to Bohr than, for example, to Einstein, a man whose deepest insights were reached in solitude.

Geo used the communal spirit he loved to gather his group for the

1938 George Washington University conference. One person he felt needed to be there was the remarkable Indian astrophysicist Subrahmanyan Chandrasekhar. Though Chandrasekhar was only twenty-seven years old at the time of the conference, Geo recognized him as one of the brightest talents in astrophysics. Eight years earlier, already fluent in the tools of modern physics, Chandrasekhar had decided to make use of the three-week steamship journey to Europe from his native India to perform a set of calculations he had recently begun thinking about. He wanted to know if the existing notions of stellar collapse needed to be revised in light of recent developments by Paul Dirac, who had incorporated relativity theory into quantum mechanics. It was believed at the time that the end point of a star's normal nuclear-fuel-burning life was a smooth transition to a so-called white dwarf star; by the time the boat docked in Europe, Chandrasekhar had realized that the new insights in theoretical physics led to a different conclusion. If the star originally had a large enough mass, now known as the Chandrasekhar limit, internal pressure could not prevent the ensuing collapse at the end of the star's life. There would be no resulting white dwarf. It wasn't clear what the alternative was, but the question could not be ignored.

Arriving in Cambridge, England, in 1930 on a fellowship for graduate work, Chandrasekhar presented his results with eager anticipation. Arthur Eddington, the dean of Cambridge astrophysicists, didn't believe they were correct; neither did his colleagues. They felt that Chandrasekhar had used the techniques of quantum mechanics in a domain where they were not applicable. Moreover, they were condescending toward the youth from the Raj. Fifty years later, Chandrasekhar's by-then universally accepted conclusion would be one of the main reasons for his receiving the 1983 Nobel Prize in Physics.

After two years of being relatively ignored by the Cambridge astrophysicists, Chandra, as everybody came to call him, went off to Bohr's Institute in Copenhagen. There he found a different atmosphere. Welcomed to the same boardinghouse that Geo and Max had lived in the year before, he acquired a group of companions who supported him, encouraged his work, and continued their loyalty to him when he went back a year later to Cambridge.

However, absorbed as they were by the emerging problems in nuclear physics, the Copenhagen group did not fully grasp Chandra's approach to stellar dynamics. Chandra remembered Bohr saying to him during that year in Copenhagen, *"Well, I've always been interested in astrophysics, but the first question I would like to know about the sun is: where does the energy come from? And since I can't answer that question, I do not think a rational theory of the stellar structure is possible."*

Chandra thought this reasoning was backward. In his opinion, you had to first understand what conditions prevailed inside stars and then ask what nuclear reactions produced the desired energy. The contrast was between a microscopic and a macroscopic approach. To use an analogy, imagine planning a suburb. Bohr's counterpart would suggest that one needed to begin with a plan for how to build a typical residence, while Chandrasekhar's would maintain that the overall size of the municipality and street patterns needed to be addressed first.

Since accumulated evidence was making it increasingly clear that hydrogen and helium were overwhelmingly more abundant than all other elements in the universe, and in the sun in particular, attention was focused on hydrogen-to-helium conversion as an energy source. A hydrogen nucleus is nothing more than a single proton, while a helium nucleus consists of two protons and two neutrons. The study of nuclear physics had advanced to the point where it was understood that under the right conditions, four hydrogen nuclei (that is, four protons) could be fused into a single helium one, with an accompanying release of considerable energy, making such a transition the preferred microscopic source powering the sun. Still, the details of how this happened and what the surrounding pressure and temperature conditions in the solar interior needed to be were far from clear.

In planning for the 1938 George Washington University conference, Geo thought that both solar dynamics and nuclear physics,

Chandrasekhar's and Bohr's approaches, should be discussed. The key was to gather young astrophysicists and nuclear physicists who could communicate with one another and overcome the barriers of specialization—to bring them together for a few days of discussions and see what they came up with. This was the same philosophy Max had adopted in approaching the Berlin biologists.

Chandra, who had moved to the University of Chicago in 1936 and had already experienced exposure to both camps, quickly agreed to attend the conference. Geo and Teller knew they should also attempt to attract the very best in both fields, even those who were not so keen on crossing the boundaries. Getting Hans Bethe to come warranted extra effort. Though he was only thirty-one years old, nobody knew more about the details of nuclear reactions than he did. Bethe finally agreed. He had attended and enjoyed the 1937 meeting, but as he later said, "I was totally uninterested in astrophysics, so I said I wouldn't come. Edward Teller persuaded me nevertheless to come, and it turned out to be probably the most important conference that I attended in my life."

The conference began on March 21, 1938, with a talk by the astrophysicist Bengt Strömgren, who presented estimates of the densities and temperatures expected in stellar interiors. Geo followed with a talk about nuclear physics. Chandra then presented his conclusions on stellar cores, and the mathematician John von Neumann, a friend of Geo's and Chandra's, added some comments regarding general relativity. It was very exciting as the speakers jumped from subject to subject, from field to field, always circling back to the central problem of what powered the sun and other stars like it. At the meeting's end, the participants issued a statement:

> Several specific contributions had been made toward formulating the next line of attack on the problems of stellar energy from the points of view of both physics and astronomy. Specific problems needing solutions were crystallized in these discussions, but an equally important

result was the stimulation of an active mutual interest on the part of the two groups represented.

This is exactly what Geo had hoped would happen: physicists and astronomers each recognizing what the others had to contribute and how moving successfully forward would require knowledge from both fields.

Afterward, Chandra, Geo, and Merle Tuve, a physicist from the Carnegie Institution, wrote a brief note, "The Problem of Stellar Energy," which was published in the May 1938 issue of *Nature*. Bethe, intrigued by the discussions at the Washington conference, decided to look for a solution to the problem they addressed there. He found a set of nuclear reactions that had not been considered before but that could also lead to a hydrogen-to-helium conversion. With a different sensitivity to stellar core temperature, these reactions turned out to be particularly important for stars with magnitudes comparable to that of the sun. Completing all the calculations in little more than a month of extraordinarily intense work, Bethe concluded that the hydrogen-to-helium conversion could explain stellar energy generation. In an interview he gave many years later, he reflected on what the discovery was like:

Well, it's the kind of satisfaction I have gotten five or six times in my life, when I really got to some important result. I can't tell you much more. I worked very feverishly. I guess I probably worked 15 hours a day or something like that, and I just couldn't leave off working on this.

In 1967 Bethe was awarded the Nobel Prize in Physics *"for his contributions to the theory of nuclear reactions, especially his discoveries concerning the energy production in stars."* It was the first time the Nobel committee officially recognized astrophysics.

Teller, who thought stellar thermonuclear reactions were, as he put it, *"Gamow's game,"* described Bethe's effort as

most remarkable. He made a systematic study of every conceivable thermonuclear reaction, catalogued all of the meager experimental data of that time, and made some marvelously enlightened guesses about nuclear reactions that had not yet been proven in experiments. His treatment was so complete that nothing could be added to his work during the next decade. Bethe proved himself the champion at Gamow's game.

Science is often this way. One person sees the problem, a second frames it, several attempt to find an answer, a group discusses how to proceed, and then an individual delivers a clear resolution to the puzzle. Those with an interest in the history of a discovery who trace the chain of events will see that credit usually goes to the one who gives the final explanation, even though this may sometimes seem unfair. In this case, though it may have been *"Gamow's game,"* Bethe is remembered as the scientist who showed how the sun generates its power.

As often happens, though, when a significant step forward is taken, a second, even bigger one is then needed. A whole mountain range may become visible only after a peak in the foreground is scaled. The new reactions Bethe had considered involved carbon and nitrogen acting as effective catalysts in the hydrogen-to-helium transformation. He had simply assumed their presence in stars, but this did not address the question of where and how these elements were created. What forces in the universe produce them, as well as iron, carbon, silicon, and all the other members of the periodic table? Explaining the life of the sun was a great triumph, but what about the life of the universe?

Geo now began to consider once again the most basic question in cosmology: How did the universe originate? Did it begin with just hydrogen? If this was the case, it seemed that only helium would be formed in stars like the sun. This meant the other elements must have arisen elsewhere, perhaps before stars such as the sun developed, but where might that have been?

Many of these puzzles were not new ones for Geo. He had been stimulated to think of them three years earlier, after being intrigued by a finding of Enrico Fermi's Rome group. They had observed that bombarding nuclei with neutrons could form new and heavier nuclei if some neutrons stuck to their targets after having been slowed down. Geo conjectured that if this could happen in the lab, it might also happen in a star. He published this idea in a 1935 issue of *the Ohio Journal of Science,* a journal even more obscure than the one in which Max, Timofeev-Ressovsky, and Zimmer had presented their thoughts on genetic mutations. This curious addition to Geo's curriculum vitae is basically only a written form of a lecture he gave at Ohio State University. As he remembered decades later, *"they paid me $100 and I thought 'My gosh, so much!' At the time I was collecting money to buy the car, so it was a big contribution."* There had been little incentive, however, to study the conditions for the buildup of elements until the question of how energy was generated in stars had been settled. Now Geo felt that it was time to begin thinking about that deep issue.

In their own way, Max and Geo were each approaching basic issues of origins. Max was beginning to study the nature of living matter, how it replicates and how it changes in doing so. Geo was pondering the aftermath of the universe's beginning. How did the material around us form?

While still thinking about these questions, Geo went back to Europe to attend a conference in Poland in June 1938. Returning to the Continent after an absence of a few years, he was immediately struck by the increased tension there. Poland's border with Russia was already sealed, and the conference's host nation provided Geo with bodyguards, fearing agents from Russia, his native country, might attempt to kidnap him. In addition, though Heisenberg had been scheduled to attend, he did not come. No Germans were present, in fact, a reflection of the growing pressure Poland was feeling from the West. Three months before the conference, in March 1938, Germany had annexed Austria during the so-called Anschluss, and in September, Hitler would sign the infamous Munich Agreement, which allowed him to march into Czechoslovakia's Sudetenland.

After the meeting, Geo stopped in Berlin. He wanted to see some of the young physicists thinking about the physics problems he was now concerned with—most particularly, Karl von Weizsäcker. Weizsäcker, only twenty-six but already a leading nuclear physicist, had been going to Copenhagen since he was twenty, brought to Bohr's Institute by Heisenberg, his close friend and mentor. While there, he had learned a good deal of astrophysics and was keenly aware of the questions Geo was pondering. He had even arrived independently at some of Bethe's conclusions about the importance of carbon and nitrogen as catalysts in the hydrogen-to-helium cycle, though not with the firmness or thoroughness that made Bethe's determinations so authoritative.

Weizsäcker had, unlike Bethe, gone on to speculate on how the heavier elements might have originated; his thoughts were not dissimilar to Geo's. He envisioned the possibility of an early universe, composed of only hydrogen, subsequently condensing and, after reaching extremes of temperature and density, exploding. He believed that nuclear reactions in that explosion might have created the heavier elements. He returned briefly to these ideas a year later, but after that his thoughts about nuclear explosions became of a much more mundane nature, as he teamed with Heisenberg on the project that would soon focus on developing a German atom bomb.

The ultranationalism, the growing militarism, and the outspoken anti-Semitism Geo saw in Berlin in 1938 frightened him. While in Berlin he would have of course made an effort to see his old friend Max, but Max had left Germany almost a year earlier, for many of the same reasons Geo had left Russia: neither of them had envisioned a very rosy future for himself in his home country.

15 Max Leaves Germany

"Mama, I see that you are still buying from Jews. I'm afraid I must tell you to choose—either the Jews or your daughter." To which my mother replied in her soft but firm voice: "My child, I choose the Jews."

Emmi Bonhoeffer, Max's sister, on family strife in the late 1930s

After his 1931 stay in Copenhagen, Geo had gone back to a very different Moscow from the one he had known a few years earlier. By the time of his return, Stalin was in undisputed control of Russia, a virtual dictator, and much of the freedom Geo valued had disappeared. Persecutions and rigged trials were rearing their head. Max would likewise soon see a Germany ruled by its own dictator, but it was already clear to him, when he went back to Berlin in the fall of 1932, that he was seeing a city that had changed dramatically from the one he had left years earlier to attend university and then embark on his foreign voyages.

By 1928 the trauma of World War I was receding and Germany had once more been fully accepted into the community of nations. It had also recovered from the disastrous burdens imposed by the Allies in the Treaty of Versailles and by the ensuing hyperinflation, but its welfare was not secure. Buffeted by crosscurrents from both the right and the left, the Weimar Republic that replaced the Kaiser's rule never had more than a tenuous hold on government. The lingering malaise from the World War I defeat and the accusations of war profiteering had fed ultranationalist groups, while many industrial workers, seeing their precarious condition continue, looked to the rise of communism as a model for a more equitable society.

The economic recovery began to slip away in early 1929, as unemployment grew. The German government had been propped up by U.S. loans, but as the financial crisis deepened across the Atlantic, these loans were recalled; Wall Street's stock market crash delivered what later came to be seen as the coup de grâce. Unable to pay the notes that came due, German industries went bankrupt, and the number of unemployed workers, 650,000 in the fall of 1928, doubled by the fall of 1929 and doubled again in the following year. The total reached more than 5 million in the fall of 1932, almost 25 percent of the workforce. The Berlin that Max saw that year was a city teetering on the edge of disaster.

The centrist parties grew increasingly weaker. Fearful of a leftist coup, industrialists joined the military in looking for support from the Nazis, Hitler's National Socialist Workers Party. Almost totally marginalized in 1928, the party had grown steadily as the economy deteriorated. By the summer of 1932, its members constituted the Reichstag's (German Parliament's) largest bloc, though not the majority. The leftist opposition forces had, if combined, almost as many seats, but the Social Democrats and the Communists were unable to join forces effectively. Hitler now challenged the aging World War I hero Field Marshal Paul von Hindenburg for the country's presidency. Hitler lost, but the election was closer than expected. Von Hindenburg, whose contempt for Hitler was palpable, resisted appointing him as chancellor, but on January 30, 1933, he capitulated after being warned that his intransigence might lead to a Communist takeover. He still hoped that Hitler might be controlled, but Hindenburg, rumored to be senile, was soon outmaneuvered.

Hitler demanded extraordinary powers after the February 1933 burning of the Reichstag, alleged though never proven to be the work of leftists. Those powers were granted, and the Nazis were now in control. Von Hindenburg died the following year, and Hitler let the designation of president expire, assuming full control as Führer, or "Leader," of the Third Reich.

Rampant anti-Semitism soon decimated the ranks of the science greats. Einstein, who had already struggled through a decade of protests against him because of his Jewish background and pacifist views, was in the United States when Hitler became chancellor. Immediately after the ship taking him there docked in the Belgian port of Antwerp, Einstein announced his resignation from the Prussian Academy, adding that he would not return to a Germany ruled by Nazis. A few months later, in April 1933, Hitler made use of his emergency powers to have the Reichstag pass the so-called Law for the Restoration of the Professional Civil Service, an edict aimed at eliminating Jews and socialists from government positions. Since this included universities, major departments were depleted overnight.

By the summer of 1933, Max Born and James Franck, the respective heads of the University of Göttingen's theoretical and experimental physics groups, had left Germany, as had many of the bright young talents they had nurtured. Berlin suffered as well. Within a short time most of what had made the capital such a lively beacon of the art world avant-garde was gone. The painters George Grosz and Ernst Kirchner, depicters of the city's street life, saw their work labeled as degenerate art. The team of Kurt Weill and Bertolt Brecht, who had electrified the music world with their *Threepenny Opera* and *Rise and Fall of the City of Mahagonny,* had left, as had the pioneering composer Arnold Schoenberg. The Bauhaus, leader of the modernist revolution in architecture, was dissolved in 1933; its director, Ludwig Mies van der Rohe, immigrated to the United States, joining its earlier director, Walter Gropius. Fritz Lang movies such as *Metropolis* were now a thing of the past, and photos of Peter Lorre, the star of Lang's groundbreaking *M,* would be used in a Nazi propaganda film, *The Eternal Jew.*

This was Max's new world. His father had died in 1929, and a year later his uncle von Harnack did, too. The pillars of the family were gone, and though the Delbrück, von Harnack, and Bonhoeffer family members were for the most part staunch Nazi opponents, there were some divisions in their ranks. One of Max's older sisters had married a fierce Hitler supporter. His sister Emmi, whose husband was executed by the Nazis in 1945, remembered that brother-in-law burning the liberal newspaper that Max subscribed to. On one occasion, an argument between him and Max turned violent, she recalled, with blows exchanged. Even Max's mother, a demure and proper lady, felt the need to take a stand against the political leanings of her pro-Nazi daughter. Emmi described an exchange between her mother and sister:

> After the boycott of Jewish shops on 1 April, Grünfeld called to deliver the linen that my mother had ordered. My sister rushed down the stairs and said: "Mama, I see that you are still buying from Jews. I'm afraid I must tell you to choose—either the Jews or your daughter." To which my mother replied in her soft but firm voice: "My child, I choose the Jews." This was not what my sister had expected. My mother was very definite about such things, but the conflict affected her health.

For the most part, Max tried to ignore the political situation and focus on science. When he moved back to Berlin in 1932, he had hoped his physics training might prove useful in exploring biological problems. By 1935 he was having some success in this direction, as we have seen, but it was also becoming increasingly clear to him that his chances for a career in the German university system were slim to none. The traditional path to an academic career had been to submit a collection of one's publications, thus demonstrating scholarly qualifications. You were then granted a *Habilitation* degree, which certified your appropriateness for being awarded a *venia legendi,* or permission to lecture at a university.

When Hitler became chancellor, he appointed an early adherent to the Nazi Party, Bernhard Rust, as minister for cultural affairs and, in June 1934, made him minister of science, education, and cultural affairs, the de facto head of the university system. Rust, who had decreed in 1933 that students and teachers must greet each other with the Nazi salute, now declared that the *Habilitation* thesis was only a step toward gaining approval for a teaching position. An extra hurdle was added in the form of a *"thorough and rigorous assessment of didactic abilities and especially the suitability of character and person for teaching at a university in a National Socialist State."* This was to be carried out at a special *Dozent Akademie,* where thirty-odd teaching candidates from all fields were to gather for a month or so of discussion and classes led by proven Nazis. They would decide, at the meeting's conclusion, whether a candidate had *"the suitability of character and person"* to be allowed to teach. *Suitability* was presumably a euphemism for political reliability, now important even in science. The Nazis wanted to make sure "Jewish physics" was not taught.

At the end of a monthlong session that Max attended in 1934, he was informed that he wasn't "mature enough," but nevertheless was encouraged to try again the following year. Though by then he had a better idea of what he should and shouldn't say, he never was one to bottle up his feelings, so his second stay at a *Dozent Akademie* was even less successful than the first. He again failed to prove his "suitability" and this time was not encouraged to repeat his attempt. Max, who could be brash and direct with the likes of Bohr, Pauli, and Timofeev-Ressovsky, was not likely to trim his sails to curry favor with Nazi bureaucrats.

The result was that Max would be allowed to continue his research at the Kaiser Wilhelm Institute, an autonomous semiprivate organization, but advancement in the university system seemed out of the question.

In some ways his position was more difficult than Geo's had been. Thanks to his early successes in physics, Geo had not encountered any difficulties in obtaining a university position in Russia and had also been quite sure that he would find one in the West if he managed to escape, despite the competition from the large number of refugees looking for positions. This confidence was warranted, as we have seen. On the other hand, though Max was free to leave Germany, his record of success in research was still spotty. It would not be enough to garner a faculty position outside Germany, particularly at a time when so many of his contemporaries with more outstanding publications—people such as Bethe, Bloch, Weisskopf, Teller, and Peierls—were fleeing Nazi persecution. And even though the 1935 paper with Timofeev-Ressovsky and Zimmer had already been published, it was not yet widely appreciated. The chance of Max's obtaining a position outside Germany with any promise of permanence was next to zero, and the opportunities at home seemed very limited.

At this juncture, toward the end of 1936, Max received a visit from Harry Miller, the head of the Rockefeller Foundation's Paris office. In an oral interview more than thirty years later, Max suggested somewhat offhandedly that Miller had just been *"checking up on what former Rockefeller Fellows were doing"* and that, finding him reading a book on population genetics by an English author, Miller had suggested to him, *"Don't you want to go to London and study with these people?"*

There is a backstory to this meeting. Not surprisingly, given his interest in helping young scientists and his belief in Max's judgment and abilities, Niels Bohr played a role in it. Bohr's association with the Rockefeller Foundation dated back to 1923, the year the Foundation established its overseas branch, the International Education Board, and the year Bohr paid his first visit to the United States. Having already received, among other honors, the Nobel Prize in Physics, Bohr was recognized as a commanding intellectual figure, and he was gladly granted an interview with IEB officials. This led to nearly a decade of funding for both Bohr's Institute and the numerous fellows who came to work there. However, in 1931 the Foundation decided that biology, not physics, should become its priority.

Bohr understood the rationale for the switch, and was interested in at least partially shifting the direction of his own Institute's research along this path. In January 1935 Harry Miller went to Copenhagen to see how Bohr's fledgling program was proceeding. At that time, Bohr inquired if the Foundation would support Max and his Göttingen friend Viki Weisskopf with fellowships as part of a biophysics program. This didn't work out, but it did at least bring Max once more to Miller's attention, signaling him that Max was both an individual in whom Bohr had a great deal of confidence and an excellent candidate if the Foundation wished to support physicists making the transition to biology.

When Max met with Miller in 1936, he saw the Foundation's suggestion as a way out of his career dilemmas and a chance to escape from Germany's repressive atmosphere. He reacted quickly, deciding, however, that he would rather study genetic mutations in a U.S. laboratory than population genetics in England. This was, after all, the subject that had interested him, and Timofeev-Ressovsky and he already had some research experience in the matter. In a December 17, 1936, letter to the Rockefeller Foundation, Max named the *Drosophila* laboratory of Thomas Hunt Morgan, who had by then moved from Columbia University to Caltech, as the place he was especially keen to visit. The Foundation accepted his proposal.

In asking to go to Caltech, Max was selecting one of the world's great centers of genetics research. As well as being the father of *Drosophila* studies and the mentor of many young scientists, Morgan was past president of the U.S. National Academy of Sciences and the first person to receive a Nobel Prize in Physiology or Medicine for work in genetics. Though by now seventy years old, he was still active in research and was surrounded by a very strong group, several of whom had worked with him for close to a quarter of a century. Going to Caltech would be very much in line with Max's own research plans. As emphasized earlier, he felt that the way a physicist might contribute to biology had to begin with that physicist learning the experimental data biologists were dealing with, and appreciating how they thought and how they approached problems. What better way was there to do this than going to Morgan's laboratory?

Miller and the Rockefeller Foundation responded enthusiastically to Max's proposal, on the condition that Morgan approve Max's visit, that Germany grant him an exit visa, and that the Kaiser Wilhelm Institute assure the Foundation that his position would still be available when he returned to Germany.

Lise Meitner, Max's boss at Kaiser Wilhelm, quickly acceded to this last request. Feeling increasingly threatened because of her Jewish roots, she was concerned that she might be forced out of Germany within the next year, but if she did stay, Max could have his old position back. This was a sufficient guarantee. The German government allowed Max to leave, and Morgan told the Foundation that Max would be welcomed at Caltech. The yearlong award the Foundation made to him in April 1937 consisted of a $150-a-month stipend plus travel expenses, to be used by Max *"to study the theory of mutations, with particular reference to their origin by physical agencies."* In the summer of 1937, Max paid a last visit to Bohr in Copenhagen and then was off to America. Though he certainly didn't imagine this would be the case, he would never again see his mother, his brother, or many friends and relatives. Almost ten years would elapse before he returned to Europe.

Max was never sure he had made the right moral decision by leaving. He had only look to his childhood friends and neighbors, the Bonhoeffers, to see what might have happened. There were eight children in that family, four sons and four daughters. One son was killed in World War I; the Nazis executed two others, Dietrich and Klaus, in 1945. Karl Friedrich, the oldest and the one who had helped the teenage Max pursue a career in science, went on to become a very distinguished chemist, a professor at the University of Göttingen, and an instrumental figure in rebuilding German scientific research after World War II. What would Max's fate have been? He would never know.

16 Max in the New World

I don't believe it.

Max, on hearing how bacterial viruses reproduce

Max left Germany in September 1937. Arriving in the United States, he explored New York City for a few days with old friends and then headed out to the genetics research center at Long Island's Cold Spring Harbor Laboratory. Its director, Milislav Demerec, welcomed Max and showed him his own *Drosophila* research. Demerec, who would become a good friend of Max's and an influential figure in the evolution of his career, even got him to start doing some experiments with fruit flies, but the experience was short-lived. The laboratory attracted many visitors in the summer, but by fall only a few individuals were left, and Max felt isolated and lonely. He was eager to move on to Caltech. Several years later he would return to that Long Island laboratory and love it.

The major growth of the Caltech Biology Division had begun a decade earlier, when Morgan moved there from Columbia University to establish a *Drosophila* research center on the West Coast. He brought with him several of his close associates, ensuring that the group would have a critical mass from the outset. Affectionately referred to as the Boss by all those who worked with him, Morgan welcomed Max to his laboratory in mid-October 1937, but neither he nor anybody else in his group showed much interest in the work Max had done with Timofeev-Ressovsky in Berlin. It sounded too much like physics and seemingly had only slight applicability to genetics.

Morgan suggested that Max speak to Alfred Sturtevant, Morgan's right-hand man, about research possibilities in the Fly Room. Max did so and was given a set of reprints on *Drosophila* literature. This was a sensible recommendation, but Max found it demoralizing because he realized it would take him weeks if not months merely to understand the relevant problems, much less be able to contribute anything of value. It would mean starting all over again in a field where so much was already known. Like Geo, he wanted to do the *"pioneering thing."* Discouraged, he went off with a friend on a camping trip in Arizona and New Mexico.

When he returned, he was told that while he was gone a young man named Emory Ellis had given an interesting talk on the subject of bacterial viruses. Regretting having missed the seminar, Max went down to the basement laboratory to see Ellis. As he listened to Max, Ellis heard the phrase Max would later use again and again when reacting to a new development in genetics: *"I don't believe it."* Ellis convinced him, and Max was soon enthralled. This is exactly what he had been looking for: a system for studying genetics through an organism so simple that it that did not require a great deal of equipment or previous knowledge. Bacterial viruses reproduced in hours, not days, and were so copious that sophisticated statistical techniques he was already familiar with might prove useful.

The study of bacterial viruses was then a backwater of genetics research, but in the next fifteen years, with considerable input from Max and the group around him, it would move to the very center of the field, with bacteria and viruses almost completely supplanting *Drosophila* as the focus of genetics research. The simplicity of viruses and the speed with which they reproduced introduced a new style of experiment and a new approach to genetics. When combined with insights from biochemistry and crystallography, this approach would eventually lead to an understanding of DNA, RNA, and their centrality in the replication of life.

Virus was a generic term used since the Roman era for poisons, and later for transmitters of infectious diseases. Though the identity of viruses remained a mystery, it was known by the end of the nineteenth century that some were so tiny they could pass through the unglazed porcelain filters that held back even the smallest bacteria; this class came to be known as filterable viruses.

The true pioneer in their study was a remarkable French Canadian microbiologist named Félix d'Herelle. An adventurer as well as a scientist, d'Herelle worked at different stages of his life in Guatemala making whiskey out of bananas; in Mexico and Argentina combating locust plagues with coccobacillus; in Egypt, India, and Indochina fighting plague and cholera with viruses; and in the Soviet Union trying to reform the country's health system. During this varied career he was nominated more than once for the Nobel Prize in Physiology or Medicine, was a researcher at the Pasteur Institute in Paris, and was a professor at both Yale and Leiden universities.

D'Herelle put the study of bacterial viruses on a firm footing in 1917 by clearly identifying the three stages of their development: First, the nutritional medium on a laboratory plate turns opaque after being infected with bacteria. Second, the bacteria die after being infected by the filterable viruses, turning the medium clear again. Third, the medium is passed through a porcelain filter, and the filtered viruses are set aside for a repetition of the experiment. With great intuition, d'Herelle deduced from this progression that filterable viruses were separate particles, that they penetrated the bacteria's cell wall, that they reproduced inside their host, and that they destroyed the cell structure and subsequently were dispersed in the medium that had sustained the bacteria. Adopting the Greek word *phagein,* meaning "devour," d'Herelle called the viruses bacteriophages, or simply phages.

Emory Ellis, the scientist whose work Max had gone to see, was just a month younger than Max. He had graduated from Caltech in 1930 and had stayed on at the school, receiving a Ph.D. in biochemistry in 1934. After a year's interval, Ellis obtained a fellowship from a foundation sponsoring basic research on cancer, a fellowship that allowed him a good deal of freedom in choosing his line of investigation. He decided to use the opportunity to follow up on some early work by Peyton Rous. In 1911 in his laboratory at the Rockefeller Institute, Rous had succeeded in producing cancerous tumors in chickens by injecting them with viruses. However, he had failed in his attempts to do so with mice, casting doubts on whether the results were applicable to mammals. The work was suggestive and never completely neglected, but its full import lay largely unappreciated for almost fifty years until new techniques were able to advance the study of virus-induced tumors.

D'Herelle had analyzed how viruses attack bacteria, and Rous's work had connected viruses to animal cancers. Ellis thought that studying bacteriophages might give him insights into how viruses operated and still satisfy the conditions of the fellowship he had received, awarded to him for cancer studies. Examining viruses also obviated the expenses associated with maintaining colonies of animals, making the experiments quick and simple, manageable by a single individual.

Believing that bacteria and fruit flies might employ similar mechanisms for reproducing, Max thought that seeing how viruses replicated by taking over bacterial machinery would lead to progress in genetics. His goals were clear; as he would later say, this work did not include the search for treatments for diseases: *"Such motives, noble though they are, are ulterior to our cause."* Nor did it include how it was that a phage entered a bacterial cell and its progeny broke out. He was simply interested in how phages replicated.

A great deal of knowledge regarding the chemical elements had been obtained by the formulation of the periodic table, but the deep understanding of chemical structure had only come in the wake of Bohr's 1913 model for the simplest of atoms. Might bacteriophage be for genetics what the hydrogen atom had been for quantum theory?

Since nobody at Caltech was working with bacteria, Ellis had proceeded by visiting a friend at the nearby University of Southern California, obtaining from him some samples of *E. coli*, bacteria that are now intensely studied but were then largely ignored. Ellis next visited the Los Angeles sewer department, gathering buckets of sewage, in which he found phages that attacked his bacteria.

The experimental apparatus Ellis used was simplicity itself, a refinement of d'Herelle's 1917 technique. He had a series of plates with a culture on which bacteria grew, turning the medium opaque; think of it as a lawn, the term sometimes used to describe bacterial growth. A few drops of phage were released onto the plate. Within a day a small, clear hole appeared on the lawn, indicating that the phages had devoured the bacteria surrounding the site where they were deposited. The expansion of these holes over time was then measured. This is what Ellis had reported on while Max was away on his camping trip.

Ellis was delighted to have Max join him in the laboratory. Working together over the next year, they demonstrated the details of how phage colonies grew with time, thereby fleshing out d'Herelle's conjectures. They labeled the three stages of growth as latent, rise, and saturation periods, corresponding respectively to phages penetrating the bacterial cell, reproducing, and then bursting through the cell, killing the bacterium in the process. With a rise period as short as thirty minutes, it was easy to repeat experiments and analyze the results statistically. Some prominent researchers disputed their interpretation, believing the entry of phages into a cell did nothing more than activate pre-phages already present there. Max was convinced this point of view was wrong, but he and Ellis needed time and more data to convince the skeptics.

Max had come to the United States in the fall of 1937 with the idea of staying for a year, but by the summer of 1938, deep into his research with Ellis, he was in no mood to abandon either it or Caltech. In August 1938 he and Ellis submitted their first bacteriophage paper, just in time for Max to include it in a report to the

Rockefeller Foundation, asking it to support him for a second year at Caltech. The Foundation's inquiries to Morgan about Max's progress brought enthusiastic responses. Satisfied by all this news, the Foundation was happy to renew the grant. Unfortunately, the foundation that supported Ellis now asked him to redirect his research to issues more closely linked to cancer. He reluctantly acquiesced. Max would have to continue alone.

It seemed unlikely that Max would be able to stay in the United States beyond the second year of his fellowship, but it was also looking increasingly problematic that there would be any position for him in Germany. Lise Meitner had assured the Rockefeller Foundation that Max would be rehired as her assistant at the Kaiser Wilhelm Institute, a precondition in his application the year before. However, since then, despite her fame and accomplishments, she had been forced to flee Germany. The Law for the Restoration of the Professional Civil Service of 1933 had removed Jews from government positions, but the Kaiser Wilhelm Institute, her primary employer, was not subject to that law because of its semiautonomous status. Very reluctant to close her laboratory and emigrate, even as the persecution of Jews increased, Meitner felt protected in part by having Austrian citizenship. This changed in March 1938, when Germany annexed Austria; she now became a German, immediately more vulnerable to anti-Jewish laws. She was threatened with dismissal and was denied a passport because scientists were not allowed to emigrate; her position became untenable.

On July 13, 1938, accompanied by a Dutch physicist friend, Meitner slipped into Holland without a passport at a lightly guarded border crossing. With her departure, Max's position as her assistant also disappeared, and it was unclear what would happen to him. Morgan did not have the funds at Caltech to support him, and it was unlikely that any other institution would be able to do so. Perhaps Max could return to the Kaiser Wilhelm Institute, but this would mean abandoning genetics research.

As for Meitner, after arriving in Holland, she flew to Copenhagen, where Bohr had been making quiet arrangements to find her a new post. She soon left for one in Sweden that had been created for her. Less than six months later, while walking in the woods with her physicist nephew Otto Frisch, she had a realization that would become the talk of the physics world and the central theme of the hastily reoriented Fifth George Washington University Theoretical Physics Conference that Geo was preparing to host. In a few years, the fruits of this realization would also change the world. Frisch and Meitner had foreseen the possibility of nuclear fission.

17 Fission

Bohr has gone crazy. He says the uranium nucleus splits.

Geo to Edward Teller in January 1939

The announced topic for the Fifth George Washington University Theoretical Physics Conference was "Low-Temperature Physics," the dates were January 26–28, 1939, and the all-star physics cast included Bohr, Fermi, and Bethe. Having arrived in the United States a few days earlier with the intention of spending the spring term at Princeton's Institute for Advanced Study, Niels Bohr was eager to attend. Enrico Fermi, who had gone directly to New York City after receiving the 1938 Nobel Prize in Physics and had accepted a professorship at Columbia University, was looking forward to getting to know the physicists in his new adopted country and to seeing some of his old European friends. Hans Bethe did not need to be cajoled into coming this time; the fourth conference had been surprisingly interesting and productive for him. Could the fifth possibly exceed the excitement for him of that prior one?

Late in the evening of January 25, Teller received a phone call from Geo, who had been having dinner with Bohr. His first words to Teller were *"Bohr has gone crazy. He says the uranium nucleus splits."*

The science leading up to the 1939 events in nuclear physics began almost thirty years earlier, when Meitner, then an aspiring thirty-year-old scientist, joined forces with an equally ambitious young Otto Hahn. Working together as a research team, first at the University of Berlin and later at the Kaiser Wilhelm Institute for Chemistry, Meitner, more of a physicist, and Hahn, more of a chemist, had many notable findings to their credit and had been nominated numerous times to share the Nobel Prize in either chemistry or physics. Recent years had seen them engaged in a research program of bombarding nuclear targets with slow neutrons, work similar to that conducted by Fermi's Rome group. Their last joint effort, performed in collaboration with the much younger Fritz Strassmann, had been submitted in a letter to the journal *Naturwissenschaften* on July 12, 1938, the day before Meitner fled Germany.

After that, Hahn and Strassmann carried on alone. In late fall they obtained some puzzling data while bombarding uranium nuclei with neutrons. Uranium, with atomic number 92 and atomic weight 238, was the final element in the then-known periodic table, and it was assumed that the addition of a neutron to the nucleus would precipitate a change into the unknown, the transuranics. These newly created elements were expected to be highly unstable, quickly decaying in a cascade that ended with different forms (isotopes) of radium. However, Hahn and Strassmann's experiments didn't support this conclusion. On December 19, Hahn wrote Meitner: "*Our radium isotopes act like barium.*" This seemed impossible to understand, because barium, with an atomic number of 56, has a nucleus that is less than half the size of radium's.

A few days later, Otto Frisch, then working at Bohr's Institute in Copenhagen, traveled to Sweden to spend Christmas vacation with his aunt Lise Meitner. Finding her pondering Hahn's letter, he suggested that they go for a walk in the neighboring woods while they talked about what it might mean. Frisch thought Hahn had probably made an error, mistakenly identifying the radium as barium, but Meitner said that couldn't be true; Hahn was too good a chemist.

But what could it mean that Hahn was finding barium? How could a half-size nucleus appear? The only possibility seemed to be that after absorbing a neutron, the uranium nucleus began to oscillate and finally divided, much as a large drop of liquid trembles and then splits into two pieces. The picture of a large nucleus as akin to a drop of liquid was not new; Geo had been the first to propose it a decade earlier, and many other scientists since then had modified and refined the scheme, but nobody had considered the possibility that a nucleus might split. Meitner and Frisch did some simple calculations on the spot. Much to their surprise, they found the prospect entirely plausible. Furthermore, by calculating the masses of the original and of the two fragment nuclei, and by using Einstein's relation between energy and mass, they saw that an enormous amount of energy would be released if and when the nucleus divided. It might be many millions of times as much as the typical energy freed in the rupture of a chemical bond.

Returning to Copenhagen at vacation's end, Frisch told Bohr what he and his aunt had conjectured. Bohr's reaction upon hearing the news was immediate: *"Oh but this is wonderful! This is just as it must be! Have you and Lise Meitner written a paper about it?"* Bohr was promptly reassured that they would within the next few days. He did not have a chance to see the manuscript before submission because he had to leave on a trip to the United States almost immediately after his conversation with Frisch. Afraid that they might lose their priority claim to this sensational discovery, Bohr promised, however, to remain silent until the paper appeared. I strongly suspect that the politically astute Bohr hoped the news might help his two good friends, both refugees from Nazi Germany, find positions that reflected their scientific merits. To ensure such an outcome, he wanted them to be given full credit.

Frisch and Meitner's conjecture was based on Hahn and Strassmann's result, but this was a chemistry experiment. It studied only the identities of the decay products that resulted from the absorption of a neutron by a uranium nucleus. A properly designed physics experiment could directly observe the fragments emerging as the nucleus split, not just the end products. Knowing precisely what to look for, Frisch found it easy to set up the apparatus for observation. He soon showed conclusively that the uranium nucleus was split into two fragments, barium and krypton nuclei. Asking a biology colleague for advice on what to call the process, he was told about the division of cell nuclei. Thinking it appropriate, Frisch gave his new find the name nuclear fission.

Meanwhile, on January 16, Bohr arrived in the United States not yet knowing the result of Frisch's experiment; the frugal Frisch had not wanted to spend money on a telegram. Bohr told Frisch he would remain silent about the discovery until he heard from him, but he forgot to transmit the agreement to his traveling companion Léon Rosenfeld, a Belgian physicist Bohr had worked closely with for many years. On arrival, Bohr and his son stayed in New York, but Rosenfeld went straight to Princeton. Unaware of Bohr's promise, Rosenfeld gave an informal seminar that very evening, during the course of which he discussed the work of Hahn and Strassmann and the ideas of Frisch and Meitner. The effect was electric: the physicists in attendance rushed to spread the news.

Arriving in Princeton the next day, Bohr was chagrined to discover from Rosenfeld what had transpired. Three days later he submitted to *Nature* a short letter; its main purpose was to ensure that Frisch and Meitner were given full credit for the discovery. In any case, on January 20, the day he sent the letter off, the German journal with Hahn and Strassmann's paper arrived; the cat was now out of the bag. Others, most particularly Fermi's new Columbia University group, began to plan an experiment similar to Frisch's. Though Frisch's experiment had still not been reported in the United States, Fermi knew exactly what he needed to do, as he and his Rome group would have observed the fission fragments in their own earlier experiments had they not, for other reasons, placed an absorbing shield around their uranium target.

With this as a background, the fifth George Washington University conference was convened on the afternoon of Thursday, January 26. Geo welcomed the participants and then announced that although the declared topic was "Low-Temperature Physics," there would be special presentations on another subject. Sitting in the meeting's front row, Bohr rose to his feet and began in his usual soft voice to explain the Hahn-Strassmann experiment and its interpretation by Frisch and Meitner. Geo next called on Fermi, who went over the same material and then discussed the experiment that would test the hypothesis, the very experiment that Frisch, unbeknownst to Fermi, had already carried out two weeks earlier.

The conference was buzzing with the news. Merle Tuve, an experimentalist at the Carnegie Institution's Department of Terrestrial Magnetism, and the conference's co-sponsor, remembered saying right away to his co-worker Larry Hafstad, *"Let's run this experiment tonight."* They slipped out of the auditorium and set to work. By Saturday morning they were able to invite the Thursday speakers to see fission fragments being produced. A few days later, to his great relief, Bohr received the articles by Frisch and Meitner. Their claim to priority was safe.

The news regarding fission opened up a new possibility that physicists found very alarming: the idea of using uranium to build an extraordinarily powerful weapon. I suspect nobody grasped all the implications as quickly as Leo Szilard, a brilliant Hungarian-born physicist who had come to Columbia University from England almost exactly a year before the conference. He had already realized in September 1933 that a weapon of this sort could be built if one could find a nucleus that released a great deal of energy after absorbing a neutron, provided that two or more neutrons were freed immediately after absorption. The idea was that one neutron would disappear in each nuclear interaction but at least two would be produced. The result would be a chain reaction, a rapidly increasing number of neutrons, energy released every time one struck a nucleus. If the whole process took place quickly, a small block of the right material could form an extraordinarily powerful bomb. Of course, if the chain reaction was too slow, the block would blow apart before the full reaction set in, and the bomb would fizzle.

The entire possibility was ignored at the time—too speculative, too remote. However, as soon as Szilard heard the news about fission, he knew how close to reality his vision had been. On January 25, the day before the conference started, he wrote to Lewis Strauss, then a New York financier interested in nuclear power and much later chair of the U.S. Atomic Energy Commission:

> *The Department of Physics at Princeton University, where I have spent the last few days, is like a stirred-up ant heap.*

Later in the same letter, he went on to discuss how these new developments

> *might lead to a large-scale production of energy and radioactive elements, unfortunately also perhaps to atomic bombs. The new discovery revives all the hopes and fears in this respect which I had in 1934 and 1935 and which I have as good as abandoned in the course of the last two years.*

The unfolding fission story still had one more big surprise in store for physicists. In early February, less than two weeks after the conference had ended, Bohr met George Placzek, a European refugee now on his way to a professorship at Cornell University. Placzek, an expert in neutron scattering, had worked with Bohr in Copenhagen, so conversation between them was very easy. He told Bohr that he found some features of the fission experiments puzzling: uranium that underwent fission seemed to behave in certain ways differently from ordinary uranium. Bohr was also perplexed but then realized what the answer had to be: only a rare form of uranium, U-235, undergoes fission when struck by slow neutrons. Though common uranium, extracted from ores, contains a small admixture of this rare form, it was enough to show up in the experiments Hahn and Strassmann conducted.

Bohr found this conclusion, that only U-235 underwent fission, reassuring, for it meant that the chain reaction Szilard envisioned would not be possible unless large quantities of the rare U-235 could be obtained. Since U-235 and U-238 are chemically identical, ordinary means of chemical separation would not work. Enriching the uranium would therefore require a massive industrial effort, and Bohr thought no nation would be able to embark on such a quest. Others feared that in the event of war, Germany would not shrink from attempting to develop such a weapon, and the consequences of its succeeding were so frightening that the Allies would also have to undertake the endeavor.

During 1939 the likelihood of a war was increasing, and discussions about the mounting tensions in Europe by the new refugees were hardly reassuring. Physicists began to realize that, one way or another, their future research was likely to be intertwined with military needs and that even nuclear physics might play a role in the unfolding drama. During the summer of 1939, trying once more to think ahead, Szilard began to worry about the uranium in the Belgian Congo, which had the world's largest known supplies of the ores. Since Germany might easily occupy Belgium in the event of a war, he was afraid the mines might fall into German hands.

Szilard thought that a warning to the U.S. government from a relative unknown would carry little weight, but Einstein's prestige was such that he could write directly to the president and be sure that the letter was read. Since Szilard and Einstein were old friends—they had worked together in Berlin and even taken out joint patent applications—he didn't hesitate to approach Einstein. Szilard, who had no car, got his fellow Hungarian, Teller, to drive him out to Long Island, where Einstein was vacationing, and briefed his friend on the developments in nuclear fission. Einstein had been unaware of them, but quickly grasped the potential danger. On August 2, 1939, he sent President Roosevelt a letter Szilard had drafted for him. In it he informed the president that it was *"almost certain"* that a nuclear chain reaction could be obtained using uranium. This would generate *"vast amounts of power"* and furthermore *"it is conceivable—though much less certain—that extremely powerful bombs of a new type may thus be constructed."* This letter is often taken as the symbolic beginning of U.S. efforts to build an atom bomb.

Within a few years one immense industrial attempt to separate uranium-235 from the more conventional form was set up in Oak Ridge, Tennessee. Furthermore, under J. Robert Oppenheimer's scientific leadership, many of America's best scientists were gathering on a mesa in New Mexico to plan the assembly of the bomb. Given Geo's expertise with nuclear reactions, you might have expected that he would have been asked to join the physicists at Los Alamos, but this didn't happen. As he says in his autobiography, *"It would have been, of course, natural for me to work on nuclear explosions, but I was not cleared for such work until 1948, after Hiroshima."* Geo attributed his exclusion to having been a colonel in the Red Army, the rank awarded more than twenty years earlier so he could be paid commensurately for teaching physics to military cadets. This seems an insufficient reason, since so many "enemy aliens" worked at Los Alamos. Moreover, George Kistiakowsky, the Bomb Implosion Division head, had also been an officer in the Russian Army and had even seen combat; of course, it had been the White Russian Army, perhaps more acceptable given the developing paranoia about Russian intents. I suspect Geo was not asked to join the team at Los Alamos because of his well-deserved reputation for being a scientific free spirit. Los Alamos was looking for brilliance, but brilliance with discipline, for scientists who were known to be hardworking, thorough, and reliable. Oppenheimer was having enough trouble handling Teller; adding Geo to the mix . might have been too much.

18 Supernovae and Neutron Stars

The Urca Process results in a rapid disappearance of thermal energy
from the interior of a star, similar to the rapid disappearance of
money from the pockets of the gamblers at the Casino da Urca.

Geo, on his work on neutrinos emitted by supernovae

During that summer of 1939, while Szilard in New York City
was worrying about the explosion of uranium because of nuclear fission, Geo in Rio de Janeiro was contemplating the possibility of whole
stars exploding. If this occurred, the amount of energy released in seconds would be comparable to the sun's output during its entire life.
As usual for Geo, his research was marked by his unmistakable style
of interspersing serious science with serious fun.

Geo, Rho, and three-year-old Igor had gone off to Rio in the summer of 1939. They lived in Urca, Rio's fashionable district. Dominated by the Sugarloaf Mountain, it also housed Rio's most famous
nightclub, the Casino da Urca, known for its gambling facilities and
fabulous floor shows. This is where, that very summer, the Broadway
impresario Lee Shubert saw Carmen Miranda dancing the samba in
platform shoes with piles of fruit balanced in a turban on her head. He
signed her to a contract; within a year the "Brazilian Bombshell" was a
star of stage and screen, the highest paid woman in the United States.

While in Rio, Geo made friends with a young Brazilian astrophysicist named Mario Schönberg, helped him get the necessary funds to come to Washington for a year, and began talking with him about collapsing stars. Their collaboration led to the discovery of a class of nuclear reactions that play a crucial role in stellar dynamics. It was their version of a Brazilian Bombshell. Astrophysicists know its action by the name Geo gave it: the Urca process, making it almost surely the only scientific process named after a nightclub.

Doing his customary *"pioneering thing,"* Geo was thinking far ahead of most of his contemporaries. It is now commonplace for physicists trained in the field of elementary particles to look for interesting applications of their ideas in astrophysics, but at the time it was a novelty. Working on those subjects myself, I frequently consult Georg Raffelt's 1996 textbook *Stars as Laboratories for Fundamental Physics*. Its first paragraph reads:

> *More than a half-century ago, Gamow and Schoenberg (1940, 1941) ushered in the advent of particle astrophysics when they speculated that neutrinos may play an important part in stellar evolution, particularly in the collapse of evolved stars. Such a hypothesis was quite bold for the time because neutrinos, which had been proposed by Pauli in 1930, were not directly detected until 1954. That their existence was far from being an established belief is illustrated by Bethe's (1939) complete silence about them in his seminal paper on the solar nuclear fusion chains.*

The Urca process is vintage Gamow, in which disparate insights from many areas of physics and astronomy are applied to solving an important problem. It doesn't exactly explain why stars explode as supernovae, but elements of the process provide the key to the solution that was found decades later. As happened so many times, Geo's work was far ahead of anybody else's. In fact, others hadn't even realized there was a problem to be solved. As we shall soon see, the same sequence occurred a little later, when Geo began to think of an even bigger explosion, the Big Bang that created our universe.

The story of how the Urca process was discovered begins in the early 1930s at Caltech. At that time an odd pair was often seen walking around campus, conversing excitedly in German. Walter Baade, the elder of the two but not by much, was a highly respected astronomer who had been lured to California by the superb observational facilities available at nearby Mount Wilson. Fritz Zwicky, his companion, was an excitable, famously difficult Swiss theoretical physicist whose interests were increasingly veering toward astrophysics.

Their conversation often centered on the topic of novae, stars that without prior warning grow much brighter, by a factor of ten thousand or more. In some cases, before their sudden appearance in the night sky, they are at first so faint as not to be even visible. Following their debut, novae fade back to normal brightness over the course of a month or more. The only explanation compatible with known theories was that of a stellar explosion triggered by a rapid change in the star's makeup, but there was little understanding of what that change might be.

Novae were puzzling, but there was an even stranger matter on Baade and Zwicky's minds in 1932: some of the novae seemed unusually bright. As Baade collected better distance measurements, he realized that these novae were a great deal farther away than previously thought. This meant they had to be enormously brighter than common novae. He and Zwicky called this new type of star a supernova.

The conclusion was astonishing. There was still no satisfactory explanation of the mechanisms generating the sun's energy, and here was something a billion times more powerful. Zwicky soon came up with a possible solution to the puzzle. Ordinarily the nuclear furnace in the stellar interior generates an outward pressure that balances the inward pull of gravity. This keeps a star's size roughly constant for millions or even billions of years. When the fuel in the core is exhausted, however, the whole star will cave in. Could this cause a sudden burst of light? Might this be a supernova?

The change in gravitational energy accompanying the collapse would be large enough to produce a supernova explosion only if the remnant star was no more than a few miles across. This seemed impossible, but in mid-1932, Zwicky came up with another brilliant idea. The neutron had been discovered a few months earlier. What if the collapse's end point was a compact sphere of neutrons? Neutrons, unlike protons, have no electric charge, so they don't repel one another; a mass comparable to a star's could fit within a radius of a few miles if it was made up entirely of neutrons. This seemed to kill two birds with one stone. The ball of neutrons provided a natural end point for the collapse, and the energy generated would be enough to produce a supernova provided the collapse was quick enough. Zwicky and Baade gave their creation the name neutron star. Their hypothesis had an additional attractive feature. As previously discussed, Chandrasekhar had concluded as early as 1930 that a large enough star would collapse at the end of its life, but how might that collapse end? Here was a possible explanation.

Astronomers found Baade and Zwicky's distinction between novae and supernovae interesting and credible; it was based on careful and reliable observations. However, they greeted the notion of a collapsing star ending up as a ball of neutrons with great skepticism. The idea of a star with a radius no larger than a good-size city seemed to them to be science fiction.

Geo didn't think it was science fiction, but he did realize in 1939 that all the discussions about neutron stars and supernovae had missed a crucial point. Zwicky's picture of an inward-falling star becoming a supernova had a fatal flaw. The vast amount of energy produced during the collapse would generate so much heat that the star's outer layers would be pushed back. The final collapse would take years instead of minutes. It was like the difference between a rock falling through air or through molasses. In one case it lands with a bang, and in the other with a gentle slide. If something didn't quickly remove energy during the collapse, the observed burst would not occur. Using the recently developed analogy to atomic weapons, this would be the stellar equivalent of a bomb that fizzled.

On April 1, 1941 (not surprisingly, April Fools' Day was Geo's favorite publication date), Geo and Mario Schönberg's article "Neutrino Theory of Stellar Collapse" appeared in *Physical Review*. In it they proposed a way in which energy could be quickly dissipated by the collapsing star. If electrons and protons in the star fused to form neutrons and neutrinos, the neutrinos would escape in seconds, cooling the core enough to cause a rapid collapse and leaving the neutrons as the new core's constituents.

This is when the Urca process was officially born. As Geo said in his autobiography, *"the Urca Process results in a rapid disappearance of thermal energy from the interior of a star, similar to the rapid disappearance of money from the pockets of the gamblers at the Casino da Urca."*

Geo had hoped that *Physical Review,* known for disliking jokes in its publications, wouldn't reject the article because he had used the name Urca. In case of trouble with the editors, he was prepared to offer the explanation that Urca stood for "Unrecordable Cooling Agent." To his relief, he wasn't asked to explain his choice of words. The name has stuck; the Casino da Urca has long shuttered its doors, but the Urca process continues to be discussed by astrophysicists.

The details of most of those 1930s speculations by Baade, Zwicky, Schönberg, Geo, and others have needed to be modified, but it is still surprising to see how essentially correct they are. Supernovae are indeed different from novae, and they are accompanied by the creation of a neutron star. Perhaps even more surprising, neutrinos emitted during the creation of a supernova have actually been observed. I am quite sure that no physicist and certainly no astronomer in 1941 dreamed that within a few decades this would become possible, but science advances rapidly and in unforeseen ways. Half a century after Gamow and Schönberg's paper, Ray Davis and Masatoshi Koshiba were on their way to Stockholm to share the 2002 Nobel Prize in Physics *"for pioneering contributions to astrophysics, in particular for the detection of cosmic neutrinos."*

19   Max Meets Manny and Sal

Between bacteriophage and myself it was love at first sight.

Salvador Luria, on first observing bacterial viruses

In August 1939, while Szilard in New York was worrying about nuclear explosions and Geo in Rio was thinking about stellar ones, Max was hard at work in his Caltech laboratory contemplating the explosive growth of bacteriophages. Had he been able to make progress in the mid-1920s on studying novae, things might have been different. Perhaps he would now have been an astronomer concerned with stellar reactions instead of a physicist turned biologist. However, by the summer of 1939, though Max was meeting with some success in his new venture, he was also becoming increasingly pessimistic about having any future in science. His Rockefeller Foundation fellowship was scheduled to end in September, and nothing else seemed to be available. Morgan, now a believer in the value of Max's enterprise, would have liked to help but didn't have the funds.

Max was soon going to be left with only the small amount of money he had managed to save during the previous two years. More than ever, he wanted to continue his work in the field of biology but knew this was probably impossible if he went back to Germany. All his doubts about a return were strengthened on September 1, 1939, when Poland was invaded and World War II began.

In June 1939 Max had asked the Rockefeller Foundation for assistance in finding employment in the United States. They had no immediate suggestions but began making discreet inquiries. Fortunately for him, the Foundation believed that his approach to biology would prove fruitful and, having already invested so much in him, they were eager to see him succeed. During that fall they heard that Nashville's Vanderbilt University might be willing to hire a physics instructor; they suggested Max be offered the position, even though this meant, as part of the negotiation, that the Foundation would have to pay at least part of his salary for two or more years. A deal was struck; Max accepted, arrived in Nashville on January 1, 1940, and began setting up a new laboratory.

Since research was still progressing slowly at the end of 1940, Max decided to take a break and travel north to Philadelphia to attend a gathering of the American Physical Society. This would give him a chance to see his friend and former mentor Wolfgang Pauli, a new arrival in the United States. Like Lise Meitner, Pauli was Viennese-born. He, too, had unwillingly become a German citizen after Austria's annexation by Germany in 1938 and, therefore, like Meitner, was now subject to the German laws that discriminated against Jews. The fact that he was only half-Jewish made little difference to the Nazi regime. However, living in Zurich rather than Berlin and having Swiss residency meant Pauli had a layer of protection that Meitner lacked, but it didn't mean he was safe. In 1940, seeing Switzerland organize defenses against a possible German invasion, Pauli began to look for a position in the United States. The Institute for Advanced Study in Princeton responded by offering him a visiting professorship, which he quickly accepted. After a tense and difficult trip through France, Spain, and Portugal, he and his wife, Paula, managed to get places on a boat steaming out of Lisbon in late August 1940.

Max had a second reason for going to Philadelphia that Christmas. He wanted to meet a young man, Salvador Luria, who he knew had interests similar to his own. Luria had been in the United States for even less time than Pauli. Born in 1912, Luria grew up in a middle-class Italian Jewish family in Turin. After obtaining a medical degree and doing some research in radiology, he decided in 1937 to go to Rome for a year of physics study, with the vague idea that it might help his radiology research. The twenty-five-year-old Luria now had a lucky break: Fermi's group at the University of Rome took him under their wing. In his autobiography he remembers how *"that year among physicists proved to be the critical turning point in my life. It taught me to think a bit in the way physicists do, a way more analytical than the way of biologists."* It also led him to discover Max's work.

Franco Rasetti, a senior member of the Fermi group, urged Luria to examine closely an obscure paper by a young physicist whom Rasetti had met while visiting Meitner's lab in Berlin. That young physicist was Max Delbrück, and the paper was the one Max had written with Timofeev-Ressovsky and Zimmer in 1935. Studying it, Luria felt that the picture they proposed of the gene as a complex molecule was extraordinarily interesting. Even though he was unsure about how to proceed, he decided then and there that this was the field he wanted to work in. A few weeks later, he saw how he might begin. When the Rome trolley he was taking to work was stopped because of a power outage, Luria started chatting with a similarly stranded Rome bacteriologist whom he knew slightly, Geo Rita. Intrigued by the exchange they had, Luria decided to accompany Rita to his lab and have a look at the samples of Tiber water Rita was collecting in order to study dysentery bacilli. In the process, Rita was also examining the growth of organisms that attacked the bacteria. Luria had never heard of them; they were called bacteriophages.

Luria recounts in his autobiography how *"Between bacteriophage and myself it was love at first sight."* It was the same feeling Max had reported upon seeing Emory Ellis's work. Separated by six thousand miles, Max and Luria had almost simultaneously come to the same conclusion.

Luria began doing some simple experiments and was *"for the first time in my life getting excited about research."* His conviction that this was the right path for him increased upon his learning that Delbrück was also working with bacteriophage; he realized with a shock that he and Max were independently moving along the same track. Luria now applied for an Italian government fellowship, hoping to use it to go to Pasadena and collaborate with his alter ego. Unfortunately, it came up for consideration in the summer of 1938. Mussolini had initially rejected Hitler's racial doctrines, but increasingly aligning himself with his Nazi partner, Mussolini had now changed his position. In July 1938 he issued a "Racial Manifesto," depriving Italian Jews of citizenship and imposing sanctions on them; among these was their exclusion from the type of award Luria was seeking.

By the late fall of 1938, Fermi had moved to the United States—Fermi's wife was Jewish and therefore so were their children according to the new laws. The Rome physics group was dissolving. Seeing no chance of a career in Italy, Luria left for Paris, hoping to escape from fascism by emigrating. He managed to obtain a French fellowship that enabled him, at least for a while, to study the effects of radiation on bacteriophage, but he soon discovered there was no safety in Paris, either. As German troops entered the city, he fled on a bicycle. His destination was Marseilles, with the single functioning American consulate in France. Luria hoped to obtain a visa for the United States, the necessary prerequisite for a French exit visa as well as for Spanish and Portuguese transit visas. With a great deal of luck, he received all of the above, found passage on one of the last boats out of Lisbon bound for the United States, and arrived in New York City on September 12, 1940, two weeks after Pauli had.

Luria's luck continued. He managed to obtain a research position at Columbia University's medical school, and Fermi, recently installed at Columbia University, spoke to the Rockefeller Foundation about him. They agreed to provide some support. Luria also got in touch with Max, arranging to meet him in Philadelphia over Christmas. Remembering that first encounter, Luria described how Max took him to dinner with two European physicist friends, one of whom was Pauli. He had little idea what the conversation was about. *"Pauli simply asked me 'Sprechen Sie Deutsch' and without waiting for a reply proceeded to eat and speak German so prodigiously fast that I understood not a word."*

Afterward, Luria asked Max to come to New York to see his experimental facilities. Realizing how much their ideas for bacteriophage research matched, he and Max laid plans for a joint venture the following summer, at Cold Spring Harbor. In June 1941 they were briefly reunited at the Long Island research laboratory. Then Max left; he was getting married.

When Max had accepted the job at Vanderbilt, he needed to change his U.S. visa from a visitor to an immigrant one. At the time, this required a short exit from the country and a guaranteed employment position upon return. Geo and Rho had traveled to England during the fall of 1934 and, with the offer from George Washington University in hand, had obtained their new visas in London. Max decided to effect the switch by going across the Mexican border to Baja California. Stopping to see old friends in California on his way south, he went to a party where he met Mary Adeline Bruce, known to her friends as Manny. This warm, sunny, friendly young woman had grown up in Cyprus, where her father was a mining engineer; later she had attended boarding school in Beirut and college in California. Though she was more than ten years younger than Max, something clicked for both of them; Max and Manny were married a year later. In their forty years together Manny cheerfully managed most of the everyday matters: cars, taxes, travel arrangements, and child care, as well as organizing the camping trips, puppet shows, and parties that made their life such a full one. She was the one who kept abreast of the news and worldly matters, leaving Max to his science.

In the summer of 1941, when Max left Luria in Cold Spring Harbor, he went back to California, married Manny, and had a brief honeymoon. Even though their marriage always was a very happy one, she later said with a smile, *"He couldn't wait to get back to Cold Spring Harbor."* Max, about to turn thirty-five in September, was not a promising young man anymore, but he now had a wife, a job, a talented collaborator, and a research base in Cold Spring Harbor. Worries about family and friends in Germany and the war were omnipresent, but at least his personal life was settling down.

That summer Max and Luria asked seemingly simple but extremely important genetics questions. What is a gene, how does it replicate, and what causes it to mutate? Many of these matters had been studied in great detail in *Drosophila,* but the answers were still not known. However, progress was being made. Only months earlier, two young Stanford University professors had obtained a major insight into how genes guide heredity.

George Beadle, three years older than Max, was an experienced geneticist, having spent five years at Caltech doing *Drosophila* research in Morgan's Fly Room. Like Max, he felt a simpler system than that of fruit flies was needed to make progress on understanding gene functions. Teaming up with Edward Tatum, a biochemist, he began to study *Neurospora crassa,* a reddish mold that grows quickly on a mixture of sucrose, salts, and one vitamin, biotin. In order to do so, the mold also needs to produce a set of enzymes for breaking down the sucrose into the nutrients it requires.

Enzymes, a subclass of proteins, are a necessity for living organisms. Some of them speed up (the technical term is *catalyze*) chemical reactions by bringing the reagents together; others act by breaking large molecules into smaller ones that can be absorbed. For example, amylases in our saliva break starches into sugars. Lipases, primarily produced in the pancreas, split fats into fatty acids, while pepsin, trypsin, and peptidases act in our stomachs to aid digestion. Each of these agents carries out a specific chemical function, a characteristic of enzymes.

The question Beadle and Tatum were studying, a very important one, was what in an organism's genetic makeup leads to the creation of any given enzyme? Their experiment proceeded by exposing the mold to ultraviolet or X rays, known to produce a single genetic mutation. Through careful studies they demonstrated that each mutated variant of the mold lacked a particular enzyme, not always the same one, and that the mold would not grow until that one enzyme was restored to the mix. In other words, they were showing that a single gene's mutation caused one enzyme to be missing. Their result was summarized in the pithy aphorism "one gene, one enzyme."

This conclusion forged an important link between genetics and biochemistry. Beadle and Tatum's experiment did not explain, however, what in the genetic makeup specified all the proteins that were not enzymes. Beyond that, a whole series of fundamental questions remained unanswered. What was a gene's chemical structure? How did it replicate? How was the connection between genes and proteins made? Though all these questions remained unanswered, many regard "one gene, one enzyme" as the beginning of what has come to be known as molecular biology.

Beadle and Tatum had made progress by focusing on enzymes and then following a distinguished scientific path: ask the right question and choose the right system to look for the answer. However, enzymes are only one kind of protein; others are responsible for a variety of different cellular functions, such as signaling or structure building. Proteins are large molecules (made up of carbon, hydrogen, nitrogen, oxygen, and a little bit of sulfur and phosphorus) arranged in long folded-up chains of smaller molecules known as amino acids. Since amino acids come in twenty different varieties, proteins may be thought of as hundreds or even thousands of beads joined end to end, beads selected from a toolbox containing twenty different sections. Because it was still believed in the early 1940s that genes were proteins, the very richness of life seemed to indicate the need for an essentially infinite number of genes, a wealth that could be provided by the almost limitless possibility of joining twenty different kinds of amino acids, one after another, into a chain.

Max and Luria were not challenging this position, but they were hoping they could answer some of the questions about genes by studying the way viruses attacked bacteria. This was a risky enterprise, since it still wasn't even clear that bacteria had genes. The first step toward verifying this assumption was finding a way to study how bacteria mutated. Max and Luria succeeded dramatically in early 1943 with a paper that would become, as the prominent molecular biologist Gunther Stent wrote, *"the standard by which all later papers on bacterial genetics were to be measured."*

20   Hitting the Jackpot

In the early days one of us would occasionally come up with a smelly skeleton or a piece of rotting driftwood that Delbrück would fling away as being unworthy of his paradise.

Thomas Anderson, on Max's influence on him

Bacteriophages, the special kind of viruses Max and Luria were studying, seemed to reproduce by taking control of the bacterial machinery and exploiting it. In some ways they acted like pirate ships navigating the seas and looking for prey. After casting grappling hooks on a rich galleon, the buccaneers streamed aboard the hapless vessel, seized control of its bounty, and then sank it. But did this analogy make any sense? Since bacteriophages are too small to be seen under a microscope, there was a great deal of room for fantasy.

As Max and Luria started their collaboration, a tool was developed that would be a great help to them: the electron microscope. The quantum theory notion that electrons could be viewed as either particles or waves led to the idea that a focused beam of electrons could be employed to develop a new kind of microscope. Moreover, since the corresponding wavelengths of electrons could be made much smaller than ones in the optical range, this new kind of microscope could also have a much greater resolution. Prototypes were built in the early 1930s, but it took almost another decade before ones were constructed that were useful for biological research—that is, instruments that had suitable resolution and did not destroy the specimen being examined by the intensity of their electron beam.

RCA, the Radio Corporation of America, had been a pioneer in the development of television during the 1930s, but the company's director of electronics research, Vladimir Zworykin, had an interest in constructing viable electron microscopes, believing they might also have commercial applications. Success was achieved in 1939, and one of these marvels, having a power of magnification of over one hundred thousand, was installed in RCA's Camden, New Jersey, laboratories. Since RCA was eager to exploit the tool's capabilities in new fields, the company decided to offer a fellowship to anyone who would make use of it for biological applications. In September 1940 they selected Thomas Anderson, a young physical chemist from Wisconsin.

As a student, Anderson had studied and admired Max's work on viruses. Reminiscing almost thirty years later about what Max's papers had meant to him, Anderson wrote:

> They formed a little green island of logic in the mud-flat of conflicting reports, groundless speculations, and heated but pointless polemics surrounding the Twort-d'Herelle phenomenon [bacteriophage].
>
> Little did I know that in a few years I would casually visit the tiny island that Delbrück was building and spend most of the next twenty-five years helping an ever increasing number of workers to enlarge and beautify it with pretty pebbles and pieces of coral. Eventually almost all of our offerings fell into place to form a beautiful arrangement, but in the early days one of us would occasionally come up with a smelly skeleton or a piece of rotting driftwood that Delbrück would fling away as being unworthy of his paradise.

Anderson's visits to *"the tiny island that Delbrück was building"* began in November 1941, when Luria came to see him inquiring if he was interested in trying to take pictures of phages. A quick deal was struck; work began on December 8, 1941, the day after the attack on Pearl Harbor that led to the U.S. entry into World War II and, incidentally, to Max's and Luria's classifications as enemy aliens. Since RCA conducted government defense work, this classification might have caused some difficulties, but Luria was given the necessary security clearance. (One can only wonder what might happen nowadays to an "enemy alien" carrying viruses into a government research laboratory.)

By March Luria and Anderson were ready to submit an article for publication. The electron microscope pictures had far greater resolution than those taken with an ordinary microscope; they showed the first clear images of phages and, even more than that, illustrated their life cycle. Smaller than previously thought, phages were tadpole-shaped creatures with round heads and long tails. They attached themselves to a bacterium's surface membrane, and then, after a half hour, the membrane burst and a new generation of phages, a hundred or so of them, streamed out from the corpse of the now-dead bacterium. The scenario Max had posed of how phages interact with their host was shown to be essentially correct, with one important difference: phages did not seem actually to enter the bacterium. They nevertheless transmitted genetic information through the membrane and instructed the bacterium to produce their replicas; how this occurred was still a puzzle.

The photographs made a big impression on the community studying viruses. Partially on the basis of this result, or so Luria believed, he was awarded a Guggenheim Fellowship; he chose to use it by going to Vanderbilt so that the two "enemy aliens" could work together.

No system is simpler for examining replication than the bacterial virus, an organism whose only aim is to produce more viruses. Destroying its host as it multiplies is a by-product. But studying genetics with the use of bacteria had problems that Max and Luria needed to face. The main one was that the standard technique for identifying a mutant by an external characteristic did not apply, nor apparently was there the possibility of crossbreeding. Fruit fly mutants differed by eye color, body shape, or other visible characteristics. What could one say about mutant bacteria?

Seemingly all one could do with bacteria was expose millions of them to an agent known to kill them—for example, a bacteriophage. The survivors were resistant mutants, but it wasn't clear whether the mutations occurred spontaneously or as an adaptation to exposure to the lethal agent. The Nobel Prize winner Sir Cyril Hinshelwood, widely considered the world's leading authority on bacterial cells, forcibly expounded the latter view. He also believed bacteria had neither chromosomes nor genes and that resistance to phage was due to a change in chemical balances rather than to a mutation. Max and Luria disagreed with Hinshelwood, but they needed proof if they were going to challenge the man Luria called the *"Goliath of physical chemistry."*

Max and Luria worked together on the problem through the fall of 1942 without a breakthrough. They sometimes felt that they were close, but the answer eluded them. Around Christmas of 1942, Luria cut short his Guggenheim Fellowship stay at Vanderbilt to take up a position at Indiana University; he had been planning to stay at Vanderbilt for the full year, but university positions were scarce and he gratefully accepted the unexpected offer from Indiana.

Three weeks later, he saw the step he and Max had been missing. It is not uncommon in scientific partnerships that two individuals struggle together, repeatedly going over the same issues, only to find the solution they want when they separate. This had been the story with the Copenhagen interpretation of quantum mechanics in early 1927; Bohr and Heisenberg had examined time and time again over the course of the previous months the problem of what constituted a measurement of a particle's position and velocity. Feeling frustrated and deciding they needed a temporary respite, Bohr went off skiing in Norway. Within a week, Heisenberg had come up with his famous uncertainty principle, the key to unlocking the puzzle.

In this case Luria played the role of Heisenberg and Max that of Bohr. The idea came to Luria on a Saturday evening while he was studying a slot machine in a game room of a Bloomington, Indiana, country club. Trying to fit into his new environment, he had gone to a faculty dance at the club. However, his mind was still on the problem he and Max had been thinking about for many months: how to prove that bacterial mutations occur spontaneously. Viewing the slot machines, he suddenly saw how to proceed.

You lose or get a small return most of the time you play a slot machine, but the machines are programmed to occasionally hit a jackpot. The biology analogy Luria imagined was that of a hundred laboratory dishes containing nutrients. Spray bacteria on each plate. A few hours later the bacteria will have multiplied through successive generations, and each plate will contain millions of bacteria. Then spray each plate with phages. Only the phage-resistant mutants will survive. If mutations are induced by the presence of phages, the plates will all look alike. If, however, mutants occur spontaneously, a plate's appearance will depend on when during the successive replications the mutation that protected the bacteria against phages took place. If the mutation occurs at the very end, the plate will look the same as if the mutation were caused by the phages' presence; only a few bacteria will survive. If, however, the mutation happens near the very beginning of the bacterial growth on a given plate, the bacteria's offspring will also be mutants, increasing in numbers as they replicate. By the time phages are sprayed, they will do very little harm to the bacteria on that mutant-rich plate. In terms of bacterial survival, this plate will be hitting a jackpot.

Luria went to his laboratory the next morning. It was Sunday; forty-eight hours later he had obtained preliminary results that supported his guess. He wrote Max immediately that the key was studying the fluctuations in the number of bacteria that survived. By return mail Max said he agreed and that he had immediately begun working out what distribution of fluctuations should be expected. The first draft of their paper on this study arrived in Bloomington on February 3, 1943.

The final version, appearing a few months later, was important for a number of reasons (aside from the key conclusion that mutations in bacteria occur spontaneously, not as a result of changes in the environment). The first was that it displayed the kind of sophistication in thinking and in statistical analysis that would be needed for advances in bacterial genetics; it showed a new way to advance the field. The second was that it brought together two different scientific communities. Until then, bacteriologists largely ignored genetics results, and vice versa. Geo and Teller had shown that real progress could be obtained if nuclear physicists collaborated with astrophysicists; Max and Luria were doing the same with bacteriologists and geneticists. It would still take a few years for these teachings to have their full impact, but the path had been set.

21   What Is Life?

Living matter, while not eluding the "laws of physics" as established up to date, is likely to involve "other laws of physics" hitherto
unknown, which however, once they have been revealed, will form
just as integral a part of this science as the former.

Erwin Schrödinger in his famous book What Is Life?

On February 5, 1943, two days after Luria received from Max
the first version of what came to be known as the fluctuation paper,
another event took place that would play an important role in the
development of molecular biology and in Max's career.

In faraway Dublin, the great physicist Erwin Schrödinger delivered
the first of a series of lectures on the topic "What Is Life?" Schrödinger
had arrived in Ireland by a roundabout route. Irritated by the ways in
which the Nazi takeover affected German universities and personal
freedoms, he had resigned his University of Berlin professorship in
1933 to take up a post at Oxford, but the unconventional and quixotic
Schrödinger had not liked the English university. He was an advocate
of free love and found the prevailing monastic atmosphere suffocat-
ing. In 1936 he resigned his Oxford professorship and returned to his
native Austria, only to be dismissed by the Germans from his new
post in Graz after their 1938 takeover of his homeland.

Schrödinger made inquiries to see if they wanted him back in
Oxford, but he was rebuffed. At that point Eamon de Valera, the head

of the new Irish republic, contacted him. De Valera had always had an interest in physics and was now trying to start an institute for advanced study, modeling it on the one in Princeton that included Einstein. An offer was made to Schrödinger, and he accepted, arriving in Dublin in October 1939, accompanied by his wife, his mistress (who, he insisted, should be considered as no different from his wife), and the child he had had with his mistress. Though officially quite puritanical, Dublin was in practice remarkably laissez-faire about the unconventional Schrödinger household arrangements.

When, in 1943, Schrödinger was asked to give a set of public lectures, he decided to focus on the question of how life comes about and sustains itself. Surprisingly, given the apparently esoteric subject, the lectures were an extraordinary success, so much so that the hall was repeatedly filled to overflowing. It seemed like a curious choice, far from Schrödinger's area of expertise, but the reasons for it became clear to those in attendance almost immediately.

A year later, a small book based on those lectures appeared. Carrying the provocative title *What Is Life?* it begins with Schrödinger readily admitting that a scientist *"is usually expected not to write on any topic of which he is not a master,"* adding, however, that the increasing specialization of science had made it impossible to reach *"unified all-embracing knowledge"* unless *"some of us should venture to embark on a synthesis of facts and theories, albeit with second-hand and incomplete knowledge of some of them—and at the risk of making fools of ourselves."* Schrödinger then launched, in clear and simple language, into a discussion of heredity and mutations.

Having read the Delbrück, Timofeev-Ressovsky, and Zimmer paper, a copy of which had been lent to him by a German friend who had also taken refuge in Ireland, Schrödinger had been impressed by Max's assertion that genes are molecules and that the quantum mechanical rules that govern molecular behavior are sufficient to explain the relative stability of genetic information. Moreover, Schrödinger found Max's interpretation of mutations as molecular rearrangements so appealing that he decided to make these arguments the book's centerpiece. Its fifth chapter, entitled "Delbrück's Model Discussed and Tested," concludes by saying,

> *there is no alternative to the molecular explanation of the hereditary substance. The physical aspect leaves no other possibility to account for its permanence. If the Delbrück picture should fail, we would have to give up further attempts.*

It's worth remembering that at this time Max was still a relatively obscure physics instructor at Vanderbilt University; now Schrödinger, the man who had revolutionized quantum physics, was hailing him as the author of a seemingly fundamental insight into genetics. This was, at the very least, a boost to Max's career. But the book also seemed to encourage faith that the methodology gained by working as a physicist might prove useful in approaching genetics problems.

What Is Life? is prescient in saying

> *Delbrück's molecular model, in its complete generality, seems to contain no hint as to how the hereditary substance works. Indeed, I do not expect that any detailed information on this question is likely to come from physics in the near future. The advance is proceeding and will, I am sure, continue to do so from biochemistry under the guidance of physiology and genetics.*

If Schrödinger had stopped there, we would hail him as having a keen eye for what the future would bring. But in the very next paragraph, he seems to be raising once more the discredited notion of laws of new physics yet to be discovered:

> *No detailed information about the functioning of the genetical mechanism can emerge from a description of its structure so general as has been given above. That is obvious. But strangely enough, there is*

just one general conclusion to be obtained from it, and that I confess, was my only motive for writing this book. From Delbrück's general picture of the hereditary substance it emerges that living matter, while not eluding the "laws of physics" as established up to date, is likely to involve "other laws of physics" hitherto unknown, which however, once they have been revealed, will form just as integral a part of this science as the former.

Schrödinger is clearly asserting his desire to find other and new laws of physics. Was this the raison d'être for Max's move to biology? That seems to be going too far. Max never made such a claim, but he did often state that he was looking in biology for something akin to the notion of complementarity that Bohr considered so central in quantum physics, the need to study phenomena in two mutually exclusive ways. Is it too much of a stretch to say this is tantamount to new laws of physics?

Schrödinger's book was in many ways out of date upon publication— he did not know of Max's later work on phage nor of Beadle and Tatum's research on *Neurospora*—but its simple, clear tone and the mystique associated with the author led to its having a great influence on young would-be scientists. James Watson remembers encountering the book as a third-year undergraduate at the University of Chicago and switching to genetics after reading it:

My change of heart was inspired not by an unforgettable teacher but a little book that appeared in 1944, What Is Life? *. . . Schrödinger struck a chord because I too was intrigued by the essence of life. . . . What sort of molecular code could be so elaborate as to convey all the multitudinous wonders of the living world?*

Watson had intended to pursue a career as a naturalist. A year after reading the book he became Luria's first graduate student at Indiana University.

Francis Crick, the other author of the Watson-Crick model of DNA, also read the book shortly after it came out. His comments about it are more skeptical than Watson's: *"it was only later that I came to see its limitations—like many physicists, he knew nothing of chemistry—but he certainly made it seem as if great things were just around the corner."* Great things were indeed emerging, but they were not quite around the corner.

22   The Phage Group Grows

Delbrück wanted to model himself on his two great teachers by combining Bohr's insights with Pauli's mordant criticism, or as he put it, by becoming God and Mephisto all in one.

Max Perutz

At a 1966 symposium honoring Max on his sixtieth birthday, Luria reminisced how

Twenty-five years ago, when Delbrück and I first met, we were probably the only two people interested in phage from the point of view of "molecular biology." Our correspondence between 1940 and 1943 dealt, more often than not, with the problem of attracting geneticists, biochemists, and cell physiologists to the dimly glimpsed green pastures of this promised land.

Max and Luria's first recruit was Alfred Hershey, a Washington University, St. Louis, bacteriologist. After a 1942 visit with him, Max described Hershey to Luria on a postcard, Max's standard form of communication in those days: *"Drinks whiskey but not tea. Simple and to the point. Likes living in a sailboat for three month, likes independence."* Max and Hershey got along very well, so much so that after meeting with him in April 1943, Max and Luria began coordinating their own research with Hershey's. We might take this as the true beginning of the "phage group," once described by Hershey as *"two enemy aliens and a social misfit."* A later member of the group, years after the three founding members had died (and more than three decades after the trio shared the Nobel Prize), described them differently:

The Phage Church, as we were sometimes called, was led by the Trinity of Delbrück, Luria and Hershey. Delbrück's status as founder and his ex cathedra manner made him the Pope, of course, and Luria was the hard-working socially sensitive priest-confessor. And Hershey was the saint.

The Phage Church's first place of worship was Cold Spring Harbor. Having enjoyed the time spent there and believing it would be an ideal location for preaching the new gospel, Max and Luria decided to conduct a short course on phage genetics in the summer of 1945. The laboratory's director, Milislav Demerec, was supportive. Max took charge of the course the first summer; it lasted nine days, featuring both lectures and laboratory sessions. Students with no previous preparation were welcome. Though learning the subject matter was very demanding, the clarity of the presentation and the intriguing possibilities for future research quickly established the program.

The course, extended the following year to three weeks, became a Cold Spring Harbor fixture, running every summer for the next twenty-six years. It was an unusual grouping, with a mix of beginning graduate students and established professors working side by side. At first there were only a handful of students, but their numbers grew steadily as word spread about the excitement in the field and the value of the course's introduction. One of the early arrivals was the brilliant polymath physicist Leo Szilard, Einstein's friend and co-worker from his Berlin days. Szilard had been the first to envision the possibility of a nuclear chain reaction and had played a key role in the development of nuclear weapons, but he had strenuously opposed dropping atom bombs on Japan. Now reluctant to conduct further physics research, he was looking to move in another direction. Aaron Novick, later a distinguished molecular biologist but then a budding young physicist, remembered how the transition began:

> *One spring evening in 1947, as we were leaving a meeting of the Atomic Scientists of Chicago, Szilard approached me and asked whether I would care to join him in an adventure into biology. Despite his caution to think his proposition over carefully, I accepted immediately.*

Szilard suggested they start by taking Cold Spring Harbor's phage course, since that would give them a quick overview. A few weeks later they traveled to Long Island and enrolled. Novick later remembered what it was like:

> *In that three-week course we were given a set of clear definitions, a set of experimental techniques and the spirit of trying to clarify and understand. It seemed to us that Delbrück had created, almost single-handedly, an area in which we could work, and after the three-week course we felt ready to embark on our own without further preparation.*
>
> *It was evident to me that Szilard regarded Delbrück highly. Usually Szilard listened to people only as long as they had something to say that interested him and made sense. This meant that he sometimes turned away in the middle of a conversation. But whenever Delbrück was talking he stayed to listen.*

The atmosphere at Cold Spring Harbor helped foster exchanges and learning. The setting was very primitive, little more than an assortment of ramshackle houses and laboratory facilities. But everything was nearby, the ambience was extraordinarily informal, and scientists of all ages mixed without regard for age or rank. The tennis court surface left much to be desired (Max and his wife, Manny, loved playing),

Copenhagen, 1930: The weeklong meeting at Bohr's Institute for Theoretical Physics. Front row, left to right: Oskar Klein, Niels Bohr, Werner Heisenberg, Wolfgang Pauli, Geo, Lev Landau, Hendrik Kramers. Note the toy bugle and cannon on the front desk.

Cambridge, 1930: Geo talks with John Cockroft about Ernest Rutherford's suggestion, following a discussion with Geo, that Cockroft and Ernest Walton build an accelerator and use it to "crack the lithium nucleus."

Copenhagen, 1931: Geo giving Niels Bohr a motorcycle-riding lesson. Geo is to Bohr's right. To his left are Bohr's wife, Margrethe; Max; and Léon Rosenfeld.

Bristol, 1932: After his time in Copenhagen and Zurich, Max returned to Bristol while looking for a more permanent position back in Germany. Here he is shown with his friend Cecil Powell, the future Nobel Prize–winning physicist.

Washington, D.C., 1937: The George Washington University theoretical physics conference, the first really successful conference on the subject held in the United States. The topic that year was advances in nuclear physics. The figure at front center is Hans Bethe; to his left is Niels Bohr. In the second row, Isador Rabi is behind Bethe's left shoulder, and Geo is behind Bohr's left shoulder. Other prominent figures in the photograph are Felix Bloch, James Franck, Edward Teller, and Eugene Wigner.

Cold Spring Harbor, summer 1941: Max and Salvador Luria at the lab, beginning to plan their work together on bacterial viruses.

Nashville, early 1940s: Portrait of a young Max during his time as a physics professor at Vanderbilt University. The Rockefeller Foundation had helped him secure this position and leave Europe as World War II began.

Bethesda, early 1940s: Geo and his son, Igor, in the backyard of their house in the suburbs of Washington, D.C.

Pasadena, early 1950s: Max and Manny with their two oldest children, Jonathan and Nicola, in the backyard of their California house.

Cold Spring Harbor, 1953: Max and Sal Luria sitting on the porch soon after the discovery of the double helix model changed the course of molecular biology.

Berlin, 1954: Max, Manny, and their two oldest children looking at the bomb crater on the site of Max's childhood house in Grunewald. Shortly afterward, the site became part of an autobahn ramp.

Geo at work assembling a model of DNA after his interest had shifted from physics to biology.

Portrait of a smiling and still jovial Geo in middle age.

Washington, mid-1950s: Geo lecturing to students at the Junior Academy of Sciences at George Washington University.

Max in middle age with Manny at a California beach.

Cold Spring Harbor, 1963: Five members of the DNA Tie Club gather together at a meeting. Left to right: Francis Crick, Alexander Rich, Geo, James Watson, Melvin Calvin.

but the sea was inviting and people jumped in and out of the water as they wished, Max always swimming breaststroke as he had been taught in childhood. It became a sort of bohemian summer camp for scientists, with Max leading the pack.

In many ways, it was a new version of the atmosphere that Max and Geo had experienced fifteen years earlier in Copenhagen. Bohr had then presided at his Blegdamsvej Institute and at his Tisvilde cottage over a mixture of games and serious science discussions, with no difference between young and old, established and beginner. All that mattered was the ideas one had. This was also true for Max, but again, like Bohr, he espoused a very strong ethical core. You did not dare misbehave, at least not in your scientific work. All the young biologists who came in contact with Max at the time describe him in similar terms, as a moral force, incorruptible, a standard of integrity. The term *Zen master* appears often in their recollections, for he led by example rather than by dictate; you intuited how to act rather than enduring lectures in doing so. It seemed that Geo had re-created the structure of the Copenhagen meetings in Washington, but Max had done him one better by reinventing the whole Copenhagen spirit.

It is also true that Max, quite simply, was fun to be with just as Bohr had been. He characterized Geo as being *"a tremendous vital force,"* but this description applies just as well to him. In contrast, Luria described his own *"reluctance to enter the relation of almost intimacy with the students that such teaching generates and which Max evidently enjoyed."* The esprit, the enthusiasm with which Max approached science and life, made him immensely attractive to the young scientists. And it wasn't only the young. Luria chronicles in his autobiography the lifelong battle with depression he suffered, and Hershey had similar struggles, but Max was there to reassure them of the worthiness of their pursuits, the excitement of their findings, and the fun they would have. He was the joyous leader of the Phage Church.

I recently walked through Cold Spring Harbor Laboratory with Jim Watson. He has written about what it was like for him to be there in

1948 as a nineteen-year-old graduate student and how, *"As the summer passed on I liked Cold Spring Harbor more and more, both for its intrinsic beauty and for the honest ways in which good and bad science got sorted out."* Now, more than sixty years later, Watson talked to me about Max and Manny as we looked at photographs from those halcyon days in the Blackford Hall Reading Room, the same room where Max and others once held forth. It is very small compared to the new Grace Auditorium, which has a large portrait of Jim Watson by the Australian artist Lewis Miller on one wall, state-of-the-art audiovisual equipment, and a bronze double helix statue in front of the door. Yet, despite all the modern developments, the laboratory retains the old Blackford Hall, from whose windows one can look out on the sailboats in that tranquil inlet of Long Island Sound.

Watson has lived at Cold Spring Harbor for a good part of the past sixty years, acting as director of the laboratory, dynamic fund-raiser, and iconoclastic guiding spirit. During this time the laboratory has grown enormously, now employing a thousand people year round, but it has not forgotten its origins: a Max Delbrück Laboratory building is nestled by the side of buildings named for donors. Seeing how it has continued to expand, Watson expressed his concern to me that the laboratory was becoming *"dangerously big"*; he wants it to continue having the informality, ease, and critical spirit that allowed for the *"honest ways in which good and bad science got sorted out."* One can only hope this remains the case.

Max was the most forceful critic at Cold Spring Harbor, the one who above all others ensured that good and bad science was sorted out. His intuition on biological questions was often faulty, but he was always ready to be convinced of his errors, and his effort to insert rigor into discussions was welcomed in the growing field. Several biologists who knew Max told me that his greatest contribution to bacterial genetics was his keen intelligence and the critical attitude he adopted in seminars and in examining others' work. He felt that acceptance should be granted only after experiments designed to disprove a theory had been proposed, performed, and failed.

In dealing with Max on scientific matters you had to learn to fight back. Fuzzy formulations were not tolerated. Though Bohr was almost certainly the dominant intellectual influence in Max's life, as well as being his inspiration for creating the Copenhagen-like spirit for the phage biologists, there were ways in which Max was more like Pauli than Bohr. This was not true of his physical attributes; in his stamina and athleticism, Max resembled Bohr far more than the rotund and sedentary Pauli. But his habit of either walking out of a talk or declaring *"this is the worst seminar I have ever heard"* was definitely not the style of the equally critical but always polite Bohr. Max's common retort when presented with novel findings, *"I don't believe a word of what you are saying,"* was also vintage Pauli. The prominent Austro-British biologist Max Perutz has succinctly and, I believe, accurately identified Max's attitude toward science and its practitioners: *"Delbrück wanted to model himself on his two great teachers by combining Bohr's insights with Pauli's mordant criticism, or as he put it, by becoming God and Mephisto all in one."*

Some felt Max's directness and brusqueness could be hurtful, and many did not approve. In a review of a recent book by Jim Watson, Sydney Brenner, one of the most influential figures of modern biology, compares Watson and Delbrück:

> Unlike Delbrück, who was nearly always wrong, Jim was nearly always right in his scientific choices. And also, unlike Delbrück (who could be ruthlessly dismissive in his criticism and wounded many young people), Jim was constructive even though some of his ideas could be eccentric. Jim also knew what it was like to be a young person in a competitive field of science.

I don't agree with the implied comparison of the last sentence, since Max certainly knew what it was like to be a young person in a competitive field of science. The atmosphere he experienced in Göttingen and Copenhagen was without doubt as spirited as any a young molecular biologist could encounter. Nor do I think he was quite as hurtful as Brenner implies, though I am sure some felt wounded by his criticisms. I believe Max's stance came from a decision to voice disapproval directly and forcefully, making no attempt to account for human frailties. People knew at once where he stood, and they acted accordingly. Pauli was also known for saying, *"This is the worst seminar I ever heard,"* but you were aware that many speakers had been similarly criticized. And it is also true that there was a great well of affection among the other scientists for Max, just as there was for Pauli. Both listened, engaged with you, and, most important, treated all speakers equally. To them, *everybody* was capable of giving *"the worst seminar I ever heard."*

Max's critical stance was never more evident than in the June 1946 Cold Spring Harbor symposium, the first such meeting of the post–World War II era. Its topic, "Heredity and Microorganisms," could not have been more appropriate. Beadle and Tatum's "one gene, one enzyme" theory of how genes function was the conference's centerpiece, but Joshua Lederberg, a scientist who had turned twenty-one only a few months earlier, provided the most sensational result. Working with Tatum, Lederberg showed that bacteria could exchange genetic information, just as sweet peas and fruit flies do. Independently, Hershey had just demonstrated that phages also had this capacity. Max and a collaborator had reached the same conclusion about phages, though Max put it more circumspectly in his publication, saying simply *"the parents had got together and exchanged something."*

The combined results constituted a revolution in bacterial genetics, proving that sexual reproduction, understood as the swapping of genetic material, could take place in both bacteria and phages. It most definitely was not only a property of more highly evolved organisms. The assembled group at Cold Spring Harbor was enthusiastic about these developments, but Max remained critical. There is no transcript of the discussions, but one can get a sense of them from a letter Max wrote to Lederberg a year and a half later. It is vintage Max:

Feb. 26, 1948

Dear Lederberg

I refuse to believe in your bacterial genetics (and therefore also refuse to take an interest in details of the type you mention) as long as the kinetics of the postulated mating reaction has not been worked out. After our discussion last summer I thought you were going to do the obvious experiments to clean this up.

Sincerely yours

M. Delbrück

Lederberg's work was correct. He shared the 1958 Nobel Prize in Physiology or Medicine with Beadle and Tatum with a citation *"for his discoveries concerning genetic recombination and the organization of genetic material in bacteria,"* but Max's criticism had also been to the point. Lederberg's mechanism was not quite as general as he had initially thought. Max was perhaps being overly critical, but he had been that way with his own conclusions.

The correctness of Max's Berlin conjecture, that mutations were due to rearrangements of large molecules according to quantum mechanics rules, was becoming increasingly clear. And his decision to work on bacterial viruses was looking ever more prophetic, for it was now being proved that, contrary to earlier prevailing views, both bacteria and the viruses that attacked them contained genes. Molecular biology was emerging as an exciting discipline, and these two microscopic agents were becoming the preferred vehicles for understanding the mysteries of genetics.

23 Geo and the Universe

"If a person falls freely he will not feel his own weight." . . . This
simple thought made a deep impression on me. It impelled me
toward a theory of gravitation.

Einstein, on the path to the general theory of relativity

By early 1948, the time Max wrote that letter to Lederberg,
he had left Vanderbilt for Caltech and become, at forty, a father for
the first time. He had also been shaken to the bone by the news of
Germany's destruction in World War II, by the ensuing chaos after the
war, and by worries about the relatives and friends he had left behind.
By contrast, though not unaffected by the war years, Geo was leading
a life that had been and continued to be rather tranquil. His parents
were long dead, his only child was almost a teenager, he had no siblings, and his concerns for old friends in Russia were not new in the
wartime years. As for memories, he still loved Pushkin's "Mother Russia" but had no interest in going back to the Soviet Union—quite the
contrary. Nor did he ever have any illusions about what role his native
country might play on the world scene. His great fear was, rather,
the Soviet Union's continued expansion into the Western world. He
voiced this concern in an October 1945 letter to Bohr, while wishing
his old mentor well on his sixtieth birthday.

Over the course of forty years, Geo wrote Bohr more than a hundred
letters. They were mailed at various times from Germany, England,
Russia, Denmark, France, Holland, Italy, the United States, shipboard

on the Atlantic, and even from the Bikini atoll where nuclear tests were being conducted—the last was written post–World War II, when he had become a consultant at Los Alamos. The first letters were in German, but Geo soon switched to Danish and then to English, though he sometimes went back to Danish or to a mixture of English and Danish. His inimitable cartoons adorn many of the letters, and almost all of them feature his notoriously bad spelling, a constant in all languages except Russian. The 1945 letter is in English:

It would be so nice if the end of the war would mean the return of the peaceful life as some fifteen years ago when we have been drinking hot chokolade at one of your previous birthdays paa [at] Blegdamsvejen. But somehow I do not feel this way at the present moment, and it looks to me more as the eve of a great Deluge coming from the East which is bound to engulf the free man on the Earth. Sorry! I didn't start this letter to develop pessimistic truths and this is just the mood. . . . But it would be realy [sic] so much nicer if one could begin to work again on pure science without the heavy clouds hanging in the air.

He goes on to tell Bohr how much he would like to return to Copenhagen,

even though this means coming too close to the red line crossing the Europe. As soon as communications will come to a more normal state I think I will do that.

But Geo never quite abandoned his fear of being too close to the "red line." In that same letter to Bohr, he also describes briefly his Washington family life:

We are living here quietly out in the woods on what is now being considered the safe distance from Washington, my boy is ten years

old and knows quite a bit about atoms, nuclei and stars. Though his
spelling is just as bad as mine (heredity of course!).

In 1939 the Gamows purchased a house in Bethesda, then still a rural Washington suburb. I believe that *"safe distance"* is a reference to their being far enough away from the blast in the case of a nuclear weapon striking the city. Geo knew all too well what such a blast's effects might be.

As mentioned earlier, Geo did not work on nuclear weapons during World War II. Nevertheless, while continuing to live in Washington during this period, he did consult for the Division of High Explosives of the U.S. Navy's Bureau of Ordnance. The most interesting part of his assignment came about through his involvement with the bureau's star recruit, Albert Einstein. Having agreed to Einstein's wish that whatever work he undertook would be carried out in his Princeton home, the division asked Geo to go from Washington to Princeton every other week with documents for Einstein to examine. Geo described what would ensue when he arrived at Einstein's house:

> *Einstein would meet me in his study at home, wearing one of his famous soft sweaters, and we would go through all the proposals, one by one. He approved practically all of them, saying "Oh, yes, very interesting, very, very ingenious" and the next day the Admiral in charge of the bureau was very happy when I reported to him Einstein's comments.*

After dispatching these duties, the two of them would have lunch and then turn their minds to what they were really interested in, cosmology.

Modern cosmology begins in 1915 with Einstein's general theory of relativity. His special theory of relativity, formulated in 1905, had been based on two assumptions. The first one, seemingly quite innocent, was that if two observers—call them Max and Geo—were traveling with a uniform velocity with respect to each other, all they could establish was their relative motion. There was no absolute frame of reference for determining what it meant to be stationary. The second assumption, quite daring, was that the velocity of light was the same whether the source was at rest or in uniform motion.

The consequences of the special theory of relativity were revolutionary for physics, but Einstein also knew that they were limited. To go further, he would have to find a way to include accelerations and the forces that generated them. In particular, he needed to be able to treat gravity's effects.

His first real progress in this direction took place in November 1907. Einstein later remembered:

I was sitting in a chair in the patent office in Bern when all of a sudden a thought occurred to me: "If a person falls freely he will not feel his own weight." I was startled. This simple thought made a deep impression on me. It impelled me toward a theory of gravitation.

What Einstein had realized on that day in the Bern office was that an observer in free fall—say, someone in an elevator in which the cord to the roof had been cut—would not be aware of gravity. It would be as if no force were present, so that a pencil released from his hand would stay there instead of dropping to the floor. There would be no relative perception of gravity, since the pencil would be falling the exact same way the person was. In other words, special relativity would hold within the elevator for the observer and the pencil, as long as they were falling together. With this seemingly innocent insight, Einstein began incorporating gravity into his special theory of relativity. It took him almost ten years to reach a full theory, but that first insight remained what he later described as *"the happiest thought of my life."*

The shortest distance between two points on the surface of a sphere is not a straight line, and two lines of longitude, though they are parallel at the equator, meet at the poles. The rules of geometry formulated by Euclid do not hold on the surface of a sphere. What Einstein realized in the years following his initial insight was that although Euclidean geometry was sufficient for treating special relativity, he would have to go beyond its concepts to describe motion in the presence of forces. For a theory in which masses cause paths to curve, he needed a formulation of trajectories in non-Euclidean geometry. The great American physicist John Wheeler, a man who did so much to develop general relativity, summarized the theory's contents in a famous two-line phrase: *"Matter tells space how to curve. Space tells matter how to move."*

The mathematical formulation of this terse aphorism is a set of equations that relate the curvature of space and time to the presence of masses or, more generally, to that of energy and momentum. The theory makes many dramatic predictions, such as the emission of gravitational waves by masses undergoing acceleration and the existence of black holes, regions where the gravitational force is so strong that even light cannot escape from them. But in 1915, when Einstein formulated the equations, his first concern was quite naturally to see if they were correct.

He knew that Newtonian mechanics had proved to be extraordinarily successful, and the description of planetary motion Newton's greatest triumph. This meant that any deviations from those predictions were extraordinarily small, but were they detectable? One interesting case had been puzzling astronomers for over half a century: a small discrepancy in the parameters of Mercury's orbit from the value predicted by Newton's equations. A whole range of possible explanations had been offered, none of which had turned out to be true. Einstein now approached the problem; his equations were very difficult to solve in general, but approximate solutions could be obtained in certain situations without too much labor, and Mercury's orbit, a case in point, turned out to be the perfect testing ground. When Einstein saw that his formulation of motion due to gravitational forces explained the anomaly, he knew that he had arrived at a deep truth, a true joining of gravity and special relativity, one that opened a new window onto the meaning of both space and time.

Einstein did not need further confirmation, but a second situation that was relatively easy to treat with his new equations was the minute change in the path of a distant star's light ray as the ray transited past the sun on its way to earth. In order for the deflection to be sizable enough for detection, the path had to almost graze the sun's surface. This meant the observation had to be made during a solar eclipse, the only time the sun's interfering glare would be blotted out. The next such event was predicted to occur in 1919. When the results confirmed that the star's light was deflected by exactly the amount Einstein had predicted, even previously doubting scientists acknowledged the magnitude of his contribution. The press made him an overnight sensation, the heir apparent to Newton's mantle as the scientist who had explained the action of gravity.

Einstein realized that his general theory of relativity could be applied to an issue that had concerned Newton as well, the motion of the universe as a whole. The great English scientist, after concluding that the force of gravity would cause a finite matter-filled universe to collapse onto itself, tried to solve this problem by imagining the universe as infinite. Such an assumption led, however, to a new set of paradoxes that neither Newton nor anybody else had been able to solve satisfactorily.

The earth revolves around the sun, the sun moves about the galactic center, and so on, but it still makes sense to talk about more general motions in our universe. On the average, do all bodies in it move away from each other or toward each other? In the first case, the universe is said to be expanding, and in the second, contracting. The prevailing belief at the time Einstein was initiating his study was that neither was true: the universe was believed to be static. Einstein saw that his equations of general relativity predicted otherwise, but he also realized that an extra term, one that came to be known as the cosmological constant, could be added to the part of the equations that described the distribution of matter and energy. It did not alter the gravitational effects on falling apples or, more grandly, either the earth-sun attraction or motion within our galaxy, but it could be adjusted so that his equations predicted a static universe.

A few years later Alexander Friedmann, a Russian theoretical physicist more interested in the formal structure of Einstein's equations than in constructing a viable cosmological model, chose to ignore the assumption that the universe was static and proceeded to examine the equations in their original form—that is, with no cosmological constant. Following this line, he found previously unnoticed solutions that featured an initial expansion that might or might not be reversed, much as rockets fired into space can persist in their outward journeys or return to earth. But an expanding universe was considered a blemish sufficient to dismiss the models. Friedmann, as mentioned earlier, did not live long enough to see his results validated; he died in 1925, most likely from typhoid fever contracted during a vacation in the Crimea.

Had Friedmann lived, Geo's career might have taken a different course. Like many other would-be young scientists of his generation, particularly those with a mathematical bent, Geo had been very excited to learn about relativity. He says in his autobiography,

> The subject that fascinated me most from my early student days was Einstein's special and especially general theory of relativity, and I had quite a lot of somewhat uncoordinated knowledge in this field. What I needed most at that time was a strict mathematical foundation in the field. It just happened that Professor Alexander Alexandrovich Friedmann of the Mathematics Department announced at that time his course of lectures entitled "Mathematical Foundations

of the Theory of Relativity," and so, naturally, I landed on the bench of the classroom for the first of his lectures.

Friedmann's sudden death forced Geo to look in other research directions, and he was then caught up in the excitement of the unfolding of quantum mechanics. Twenty years later, however, he was back to thinking about the universe's expansion.

24   Gamow's Game

With rare exceptions, theorists have to take the world as it is presented to them by observers.

> Steven Weinberg, on Einstein's decision to assume the universe was static

While Geo was learning relativity from Friedmann in the mid-1920s, Georges Lemaître was studying the subject under Sir Arthur Eddington's guidance. Ten years older than Geo, Lemaître had been a Belgian soldier for almost five years in World War I before finally managing to resume his studies of mathematics and physics at the Catholic University of Louvain. Shaken by his war experience, he decided to study theology in parallel with the pursuit of his science curriculum. In 1923, having by then been ordained a priest in the Catholic Church, he left for a year's study of astronomy in Cambridge, England, and after that for a year in Cambridge, Massachusetts.

On his return to Louvain, he began reexamining Einstein's equations. Unaware of Friedmann's paper, Lemaître soon discovered its results on his own, but having astronomical expertise that the Russian lacked, he used these solutions to construct a true cosmological model, one that incorporated effects of radiation and was consistent with new data. However, he published his findings in the obscure *Annales de la Société Scientifique de Bruxelles* and made little or no effort to get the paper noticed by the interested community. This meant that at first, the work had little or no impact. But new experimental data were about to change that.

It had been known for a century that stars emit radiation at frequencies that are characteristic of their makeup. But around 1910 the astronomer Vesto Slipher, using the telescope at Arizona's Lowell Observatory, began noticing that the observed frequencies of more distant stars appeared slightly displaced toward the red end of the spectrum. One possible interpretation was that those stars were all moving away from earth. Such motion was known to cause shifts toward the red, but the displacements were small, and Slipher was reluctant to jump to such a dramatic conclusion.

At around the same time another astronomer, Henrietta Leavitt, was pioneering the study of a class of bright pulsating stars known as Cepheid variables. After graduating in 1892 from Radcliffe College, then called the Society for the Collegiate Instruction of Women, Leavitt had been hired at the Harvard College Observatory to analyze photographic plates of the night sky. In making the offer to Leavitt, the observatory's director praised her diligence: *"I should be willing to pay thirty cents an hour in view of the quality of your work, although our usual price, in such cases, is twenty-five cents an hour."* Leavitt's careful analyses over a period of many years concluded that there was a correlation between Cepheid variables' period of pulsing and their luminosity, the amount of light they emit per second. Cepheid variables now became invaluable tools for astronomers; they were believed to be "standard candles"—that is, stars whose brightness provides a way of measuring their distance from the earth.

All these insights came together in the late 1920s in the work of Edwin Hubble, the premier observer at California's newly constructed Mount Wilson Observatory, whose telescope was, at the time, the most powerful such instrument in the world. In what is still regarded as the twentieth century's most important cosmology result, Hubble combined Leavitt's insight of using Cepheid variables as standard candles and Slipher's determination of red shifts, doing all this with better and more abundant data than any astronomer had previously obtained. In 1929 he announced his conclusion: there was a linear relation between a star's distance from earth and the frequency shift toward the red of the light it emitted. Since the frequency shift varies directly with velocity, Hubble was in effect saying that a star's velocity relative to earth, v, is directly proportional to its distance, d. This is written as $v = Hd$. H is now known as Hubble's constant.

Evidence supports the common assumption that on very large distance scales, typically considerably greater than the size of a galaxy, space is both homogeneous and isotropic. In other words, no point or direction is different from any other one. A short calculation then shows that the only possible conclusion from Hubble's detection of the relation between stellar velocity and distance is that every point in the universe is moving away from every other point. In other words, space itself is expanding.

This was a dramatic deduction. Hubble, aside from being a great astronomer, was also a strikingly effective publicist. Tall, handsome, and an excellent athlete, he had been a Rhodes scholar, passed the bar exam as a lawyer, and quickly risen to the rank of major in the U.S. Army while fighting in World War I. He spent many nights during the 1920s and '30s at his lonely perch on Mount Wilson, but he also cut quite a swath in Los Angeles society, mixing easily with Hollywood celebrities. The 1937 Academy Awards ceremony began with the renowned film director Frank Capra, the academy's president, saying he wanted first to introduce the audience to the world's greatest living astronomer; Hubble was asked to rise and then was greeted by thunderous applause as spotlights converged on him.

The 1930 astronomy community's appreciation of Hubble's results was not quite as worshipful as Hollywood's, but it was significant. Sir Arthur Eddington, the dean of British astronomers, stood up at a January 10, 1930, meeting of the Royal Astronomical Society and argued forcefully against a static model of the universe.

Einstein's reaction was also swift. He had always disliked the cosmological constant, feeling it added arbitrariness to the beauty of his general relativity equations. He had only inserted it because he had presumed the universe was static. With Hubble's result, that need disappeared. In 1931 Einstein published a paper recommending doing away with the constant and adopting the Friedmann-Lemaître solutions to his equations. In a much-quoted remark, Einstein told Geo during their World War II Princeton talks that introducing the cosmological constant had been *"the biggest blunder he ever made in his life."*

Ironically, new evidence (about which I will say more later) suggests that a cosmological constant may be present, albeit with a very different value from the one Einstein was considering. I think that a remark by the Nobel laureate Steven Weinberg on the subject is pertinent:

> *I don't think that it can count against Einstein that he had assumed the universe is static. With rare exceptions, theorists have to take the world as it is presented to them by observers. The relatively low observed velocities of stars made it almost irresistible in 1917 to suppose that the universe is static.*

Abandoning the notion of a static universe had public reverberations as well. If the universe was indeed expanding, wasn't it natural to ask how it began, and what was happening before then? Given that he was a Catholic priest, the bespectacled and dignified Lemaître found himself at the center of a storm he would have liked to avoid. This would happen to him time and time again in the years to come, when his position as a member, and later as president, of the Pontifical Academy of Sciences made him appear to be the scientific spokesman for the Catholic Church's stance on cosmology. Steering carefully away from mixing science and religion, Lemaître expressed the view that

The idea that because they [the writers of the Bible] were right in
their doctrine of immortality and salvation they must also be right
on all other subjects is simply the fallacy of people who have an
incomplete understanding of why the Bible was given to us all.

Lemaître tried to avoid becoming embroiled in a religious controversy by making a distinction between *"beginning"* and *"creation,"* a
fine point some failed to grasp. But he did not shy away from conjectures about the *"beginning"* of the universe. In 1931 he proposed a
model with a so-called *"primeval atom,"* marking the universe's well-
specified beginning, followed by continual expansion, rapid after that
first instant, and at a less precipitous pace later, bringing it into agreement with Hubble's data.

Lemaître's model was the first example of what we now call a Big
Bang universe—namely, one with a specified beginning. This has even
led some to call him the father of Big Bang cosmology, though that title
is usually given to George Gamow. Why is this so, given that Geo's
work on the subject didn't start until fifteen years later? There are
many reasons, but the main one is that Lemaître discussed his notion
of a *"primeval atom"* only vaguely. Geo, on the other hand, examined
the details of what occurred during the universe's initial phase, before
stars or planets formed, even before atoms made their appearance.
He studied the nuclear reactions that took place, what effects expansion had, and the likely outcomes. Geo was combining knowledge
of cosmology, general relativity, and nuclear physics. By doing so, he
became the first to make a reasonable set of conjectures about what
really happened in the universe after the first instant.

Geo's path to this question was indirect. Though he had a good
understanding of cosmology from his early days as a Friedmann student, his primary expertise was in nuclear physics. But, as I've said
earlier, by the mid-1930s he was eager to leave that field and embark
again on the *"pioneering thing."* He saw nuclear processes in stellar
cores as a natural opening for doing just that. As Teller said, *"stellar*

thermonuclear reactions were Gamow's game." This game had a big success in 1938, when Hans Bethe showed how the sun could be powered by a nuclear cycle converting hydrogen into helium. But the problem of how all the other elements are formed was still wide open. Bethe underlined this by stating clearly in 1939 that *"no elements heavier than Helium can be built up in ordinary stars."* If they could not be built up in ordinary stars, where did they come from?

In 1946 Geo wrote the paper that warrants calling him *"the father of Big Bang cosmology."* As in much of his work, it is a mixture of some ideas that are very right and others that are very wrong. The main wrong idea is his persistence in believing that the heavy elements were produced well before stars began to form. He imagined neutrons *"coagulating into larger and larger neutral complexes which later turned into various atomic species."* Detailed calculations would eventually show that this cannot occur. It took another twenty years before it was realized that heavy elements are synthesized in the collapsing core of massive stars and then spewed into space when the stars explode as supernovae, but all that is part of another story.

The very right idea Geo advocated was quite simple: you have to use the equation of his onetime mentor Friedmann to calculate how fast the universe is expanding when conditions are as extreme as the universe's infancy; in other words, you need to tie together nuclear physics and general relativity. His 1946 paper concluded that

> *at the epoch when the mean density of the universe was a million grams per cubic centimeter, the expansion must have been proceeding at such a high rate that this density was reduced by an order of magnitude in only about one second.*

According to Geo, the early-stage universe expanded much faster than anybody had thought.

For a variety of reasons, Geo's paper did not cause much of a stir. The study of cosmology was in its infancy, primarily carried out by a handful of astronomers who were not knowledgeable about nuclear physics. The few physicists working on the subject were not convinced by Gamow's argument that an original explosion was really necessary. Furthermore, both groups seemed to feel that the universe's first seconds were so remote from anything they had ever considered that applying normal laws of physics to this situation was probably not warranted. Geo did not agree, but he was still a voice in the wilderness. He liked to take big intellectual jumps and see where they took him. Others, more cautiously, wanted a clearer picture of their possible landing before proceeding.

25   Bohr, Geo, and Max

It was, of course, true that I had never learned any chemistry or biochemistry, and just did not want to take the time to do so.

Max Delbrück

As usual, Geo was trying to build bridges from physics to other disciplines in considering what should be the focus of the upcoming George Washington University Theoretical Physics Conference. It would be the ninth: because of the war, he hadn't held one in 1943, 1944, or 1945, but in 1946 he was ready to start again. His first decision was to shift the date from the end of April to the end of October. This would allow Niels Bohr, who was scheduled to receive that fall a special honorary doctorate from Princeton University, to attend and, it was hoped, bring with him some of the old Copenhagen spirit.

But what should the topic be? Geo had just submitted his cosmology paper to *Physical Review*, but the time did not seem right to focus on either stellar processes or cosmology. He then had an interesting idea. He knew that several physicists seemed to be looking toward biology for a new direction in their research, and he also knew the subject would interest Bohr; the first conference devoted to the physics and chemistry of genetics had been held in Copenhagen in late September 1936. Its topic had been "The Mechanism of Mutation." Little had come of it then, but Geo thought the time might now be ripe for meaningful exchanges on that subject as well as more broad-ranging questions at the interface between physics and biology. The conference Geo had organized in 1938 that brought astronomers and physicists together had been a great success. Might an exchange between biologists and physicists be comparably important?

With this in mind, Geo selected as program title "The Physics of Living Matter."

He wanted the discussions to be as free-ranging as possible, hoping an informal give-and-take would help bridge the gap between physics and biology. His call struck a chord in both communities; responding to it, an extraordinary group of individuals, some thirty in all, gathered in Washington for three days. There was one person who, Geo knew, absolutely had to be included in the group and who, he was happy to see, responded positively: his old friend Max. Although still calling himself a physicist, Max was clearly working on biological problems and was becoming increasingly prominent as a visible example of someone with a foot in both camps. He had begun making the transition from physics to biology more than a decade earlier, but many others were now joining him, some disillusioned with nuclear physics after the war and others simply looking for new frontiers.

The war and the use of nuclear weapons had been a watershed for physicists. They had come to realize all too well the uses to which their research could be put and were now adjusting to this new reality, some with continued interest in defense issues and others by moving away from physics. Leo Szilard, who had been influential at every step of the bomb's development, starting with his 1933 realization of what a chain reaction might produce, had been particularly shaken by the turn of events. Together with his University of Chicago colleague James Franck, he had been the principal author of a report issued in secret in June 1945, two months prior to Hiroshima. It advocated dropping the bomb instead on an uninhabited location, with observers from around the world present. They were overruled.

In the summer of 1947, Szilard went to Cold Spring Harbor to attend Max's phage course; eventually he became an éminence grise of molecular biology. Franck, a Nobel Prize winner in physics and a former professor at the University of Göttingen, had switched his research to photosynthesis. Franck and Szilard both came to the 1946 conference.

One problem that clearly lay within the conference's purview and also tied into the recent development of nuclear weapons was the effect of radiation on living matter. The devastation caused by a nuclear blast was all too obvious, but this subtler concern tied in directly to the "physics of living matter." All of the scientists at the conference had studied the effects of radiation in one context or another and knew that it could inflict genetic damage on the unfortunate individuals who, although surviving the blast, suffered exposure to the accompanying rays. But how much damage was inflicted?

Hermann Muller, also present at the conference, probably knew more about the subject than anybody else in the world. A veteran of the original Morgan Fly Room at Columbia University, Muller had discovered in the 1920s that radiation induces mutations in simple organisms; he had been attempting to alert the world ever since then to the dangers of exposure to radiation, including the initially unrecognized peril of X-ray overuse. A surprise morning telephone call from Stockholm on the Washington conference's second day seemed like an omen. The call was for Muller: he had been awarded the 1946 Nobel Prize in Physiology or Medicine for his *"discovery of the production of mutations by X-ray irradiation."*

Muller was now at Indiana University. In fact his presence there was the main reason that young James Watson went to the university for graduate school in 1946. However, though Watson enjoyed very much Muller's lectures during his first year at Indiana, he soon began to take notice of the research being conducted by a much younger member of the faculty, Salvador Luria. Watson's memories of that time echo this attitude:

His [Muller's] work on fruit flies [Drosophila] however, seemed to me to belong more to the past than to the future and I only briefly considered doing thesis research under his supervision. I opted instead for Luria's phages, an even speedier experimental subject than Drosophila: genetic crosses of phage done one day could be analyzed the next.

Max had realized this nearly a decade earlier when he had gone to the basement laboratory at Caltech and seen Ellis's experiments with phages. The analyses with them were quick and simple, unlike the much more laborious ones that fruit flies required. Max had been well ahead of his time, but the decision he had made then was now gathering widespread acceptance. Watson, often described at that time as a young man in a hurry, felt that the immediate future of genetics lay in the directions Max and Luria were pursuing, not in those Muller was interested in.

The sessions of the 1946 George Washington University conference reflected this awareness. *Drosophila* had become a relative backwater in genetics; the excitement had moved to bacteria and viruses, shown only months earlier to possess genes. They were now at the center of the growing interest in molecular biology, and all the conferees were eager to hear the latest news about them. In line with this thinking, the first morning session was identified as "The Problem of the Gene—introduced by Dr. G. W. Beadle" and the first afternoon session as "Bacteriophage—introduced by Dr. Max Delbrück."

Beadle, most famous for the work with Tatum summarized by the "one gene, one enzyme" epithet, focused on the biochemical aspects of genetics. Delbrück's interests lay elsewhere. It was of course extremely important to know the action of genes, but he felt that such knowledge did not bring the scientist any closer to understanding how a gene is structured or how it encodes the necessary information for its workings. These were the kinds of questions Max hoped to answer by studying bacteriophages. In a 1978 Caltech interview, Max was asked about what many saw as his lack of appreciation for or, more strongly, his seeming *"hostility towards chemistry in the investigation of biological systems."* He replied:

> . . . If you say "One gene, one enzyme" then the question remained, how does the gene make the enzyme, and how does the gene make the gene, and this was in fact not answered at all by any of the biochemical approaches. So in a sense I think my reservations about the powers of biochemistry were appropriate, and if in addition I was glib and arrogant about it, then that was just a personality defect. I mean it was, of course, true that I had never learned any chemistry or biochemistry, and just did not want to take the time to do so.

The swirl of interests accompanied by wide-ranging discussions exemplified in Beadle's and Delbrück's presentations was exactly what Geo had been hoping for. Each conferee tried to fit the information into his previous knowledge base; much of what they heard was new and exciting, but everybody came away with something. The main emphasis was on biology's links to physics, but in an extraordinarily far-seeing way, John von Neumann, also present at the conference, began to consider its ties to computer science. He was looking for the defining characteristics of life even beyond what Beadle and Max were presenting. The conference's gathering of a small number of diverse brilliant individuals addressing a loosely defined topic encouraged this type of gazing into the future.

As well as being one of the twentieth century's great mathematicians, anecdotally known for having an encyclopedic memory and for being able to perform calculations at lightning speed, John von Neumann was a key contributor to the foundations of quantum mechanics, to the development of computers, to modern economics, and to game theory, as well as having been a very important member of the Manhattan Project.

He was almost certainly the first to think in a serious way about the analogy between computing machines and living organisms. Freeman Dyson, a professor at the Institute for Advanced Study in Princeton, has repeatedly emphasized his former colleague's—von Neumann was one of the institute's first appointments when it opened in 1933— contributions to the notion of what constitutes life, arguing forcefully a point first enunciated by von Neumann, that living organisms are defined by two basic features: metabolism and replication. Prompted in part by his interest in electronic computers, von Neumann had formulated this conclusion while studying cellular automata, self-replicating mathematical systems. As Dyson has written, when von Neumann's

> *ideas were taken over by the computer industry, these were given the names hardware and software. Hardware processes information; software embodies information. These two components have their exact analogues in living cells; protein is hardware and nucleic acid is software. Protein is the essential component for metabolism. Nucleic acid is the essential component for replication. Von Neumann described precisely, in abstract terms, the logical connection between the components. For a complete self-reproducing automaton, both components are essential.*

Von Neumann's separation between replication and metabolism is a useful key to understanding Max's work. Max was almost exclusively interested in the former, happy to leave metabolism to chemists. Replication could be studied without knowing much chemistry; metabolism could not. Wanting to know how phages reproduce, bacteria were for Max a gadget, little more than a tool that would help him get answers. Not surprisingly, since he was following Max's lead, Schrödinger's focus in his little book *What Is Life?* had also been on replication.

With the recognition that bacterial genetics was a very important research field and that Max was one of the leading figures in its development, several institutions began to woo him. One very tempting offer came from the United Kingdom. The famous physicist Patrick Blackett had set up a major research physics group at Manchester, the very place where Rutherford had discovered the atomic nucleus. He was now trying to build up biophysics, and Delbrück seemed to him the ideal candidate to direct the effort. Max was tempted; he had felt very much at home in England fifteen years earlier, and Manchester, with its illustrious history of innovation in science and technology, seemed an excellent place for the endeavor. In addition, the chance to go back to Europe had many appealing features, and adventurous Manny, who had grown up in Cyprus and gone to boarding school in Beirut, was not at all averse to the move.

Soon other offers came. The most exciting one arrived in the form of an unexpected phone call from George Beadle, who only a short time earlier had shared with Max the dais that first day at Geo's conference. Replacing Morgan, Beadle had left Stanford that year to become the new chair of Caltech's Biology Division and was calling to see if Max would join him there. Beadle thought that he, Max, and Linus Pauling would constitute a formidable triumvirate, one capable of mounting a concerted attack on the problems of genetics from three different angles. Beadle was proposing that though the three had started their careers respectively as an agronomist, a physicist, and a chemist, they could shape the emerging molecular biology while working side by side. This was an offer Max could hardly refuse. But great changes were also taking place in Max's personal life during late 1946, the most important being that Manny was pregnant.

26 Back to Germany

If anybody feels guilty, I feel guilty of not having stayed because I
had so many friends who I admire for having stayed, and having
tried to save what was to save, rescue it across this disaster.

Max, on having left Germany in 1937

Before moving across the country to Pasadena, there was
something Max and Manny needed to do. They had to go to Germany
to see his family. In June 1947, they set off on a freighter across the
Atlantic. Their son, Jonathan, born only weeks earlier, was left behind,
a war-torn country in chaos being no place for a baby. When they
arrived in Berlin, the city was in ruins, trains were running irregularly,
the currency had not yet stabilized, and refugees were everywhere.
It was hard to reach the old capital, now located in Germany's Soviet
Zone; once there, they found it hard to move around, because the
city was now divided into four sectors, controlled respectively by the
Soviets, the British, the French, and the Americans. What they found
in Berlin confirmed their worst fears.

In the early years of the war, Max's attitude about his homeland had
sometimes confused his friends by appearing to be less anti-German
than they expected. At the end of 1940, Pauli wrote a long letter to
Viki Weisskopf, their common friend. It concludes:

I want to add something about Max Delbrück. I still feel a strong tie of friendship to him, as often happens with two difficult personalities . . . even with the best of intentions, I simply cannot follow his reasoning. You would surely have thought that it would not be hard to answer whether or not it was desirable that Germany lose the war. This should not be a difficult question requiring reflection and brooding, but that doesn't seem to be the case for him. Deep in his Prussian soul, despite opposition to Nazi ideology and unlimited sympathy for Jews, the question of whether the world would be better off with German dominance still seems to be an open debate for him. It was a difficult moment between the two of us but I want to be patient with him because I like him so much.

It apparently wasn't that easy for Max to shed the beliefs instilled in him by generations of Prussians devoted to serving the state and maintaining the ideals of a German worldview. But with the war over, Max saw that the entourage of his childhood, the Delbrück, von Harnack, and Bonhoeffer families, had been decimated. Some family members had been victims of Nazi bestiality, and others simple casualties of war's horrors. They had tried to maintain the ideals of their upbringing and had paid the price.

Max was the youngest of seven siblings, three boys and four girls. Waldemar, his oldest brother, had died in action during World War I. His other brother, Justus, imprisoned by the Nazis for resistance to the regime, had escaped at the end of the war but was seized by the Russians, ostensibly to testify against Nazis but actually put in a prisoners' camp. Though the family did not hear of his death until the end of 1946, he succumbed in a diphtheria epidemic after only a few months in the camp. All four of Max's sisters survived the war, though Emmi, only one year older than Max, suffered the loss of her husband, Klaus Bonhoeffer, Justus's best friend. Klaus Bonhoeffer was also the older brother of Max and Emmi's childhood playmates Sabine and Dietrich.

The Nazis had executed both Klaus and his brother Dietrich for their involvement in a plot against Hitler.

Max's mother died of natural causes in 1943 without ever seeing Max again after his departure for the United States, but perhaps she was consoled by the thought that he was safe. Until her death she had lived in the old family house in Grunewald; the time of her death may have been fortunate, since a few weeks later a bomb completely destroyed her bedroom. The house was damaged but lasted a few weeks more, until a second bomb completed its destruction, even uprooting the old oak trees in the yard, whose branches Max and his siblings used to play on. After the second bombing, the family took what they could salvage to Emmi's house, but on April 23, 1945, that home was hit by a shell that struck a gas line. The house quickly went up in flames, leaving only ashes behind.

At the time, Emmi was in Berlin anticipating the release of her husband, imprisoned by the Nazis, but they executed him the very same day the house burned to the ground. With all hopes dashed, she left on foot to reach her children, placed weeks earlier with friends in a safe country house. She faced a two-week journey that included swimming across a frigid Elbe River with *"nothing except two hundred marks and a copy of my husband's farewell letter, both tied to my chest in a small rubber container. In case I didn't get through alive I had left the original of the letter with my father-in-law."*

The suffering did not end with the war's conclusion. In its aftermath Emmi wrote, *"I lived with my three children in an attic apartment no larger than 150 square feet. In the early months we only had one bowl, which was used for everything—cleaning herring, washing potatoes, washing clothes, preparing a meal."* Max and Manny had done what little they could by sending money and care packages to alleviate the suffering, and they would continue to do so in the ensuing years.

It is hard to imagine what Max's mixture of grief and guilt must have been. Had he been right to leave? As he saw it, there was no choice if you were Jewish, because you would have been dismissed from any position you held and directly threatened, but what should you have done if you were not Jewish? In a long interview given many years later, Max expressed his ambivalence:

> *If anybody feels guilty, I feel guilty of not having stayed because I had so many friends who I admire for having stayed, and having tried to save what was to save, rescue it across this disaster. I have seen many of those: Karl Friedrich Bonhoeffer was one of them, Hans Kopfermann was another one, and many others for whom I have the greatest admiration—Von Laue, Heisenberg too; Otto Hahn certainly.*

Lise Meitner, the director of the Berlin laboratory in which Max had worked before the war, was not burdened by his feelings of guilt. During the summer of 1945, in the wake of hearing what had gone on in the concentration camps, she wrote Hahn a long letter from Sweden, where she had taken refuge in 1938. He had been her collaborator and close friend for almost thirty years, but she now felt sickened by his complaints of how Germans were suffering:

> *. . . You all worked for Nazi Germany and you did not even try passive resistance. Granted, to absolve your consciences, you helped some oppressed person here and there, but millions of innocent people were murdered and there was no protest. I must write this to you, as so much depends on what you have permitted to take place.*
> *. . . Perhaps you will remember that when I was still in Germany (and now I know that it was not only stupid, but very wrong that I did not leave at once) I often said to you "As long as only we (the Jewish people) have sleepless nights and not you, things will not get better in Germany." But you had no sleepless nights, you did not want to see, it was too uncomfortable. I could give you many large and small examples. I beg you to believe me that everything I write here is an attempt to help you.*

Max's feelings were clearly not the same as hers. They were complicated by the traces of what Pauli had called Max's *"Prussian soul"* and would remain so even after he went back to the United States. Hans Bethe felt much less conflict. When the great theoretical physicist Arnold Sommerfeld, who had trained him (as well as Pauli, Heisenberg, and others), retired from his University of Munich professorship in 1947, he asked Bethe to succeed him and help keep the tradition alive. Bethe wrote back a warm but clear refusal, saying:

And I hope, dear Sommerfeld, that you will understand: Understand what I love in America and that I owe America much gratitude (disregarding the fact that I like it here). Understand what shadows lie between myself and Germany.

The *"shadows"* he was referring to were all too clear, but any doubts as to what Bethe meant are dispelled by an earlier phrase in the letter: *"The students of 1933 did not want to hear theoretical physics from me,"* a delicate way of reminding Sommerfeld that he, Bethe, was Jewish and not welcome in Germany after Hitler came to power in 1933. Nor did Bethe now want to go back to the country where a decade earlier a campaign had been launched against *"Jewish physics,"* epitomized by Einstein's theory of relativity.

Max did not have the same *"shadows,"* nor was he being offered a prestigious professorship (that would come later), but he also did not want to return to his homeland on a permanent basis, at least not then. Though he did try in 1947 to inform a new generation of German scientists about developments in biology, he was above all eager to go to Caltech and start research in this new environment.

The visit to Germany certainly left its mark on him. Max and Manny's son, Jonathan, thinks that the devastation they saw in 1947 *"may have influenced them to choose a lifestyle which kept very close ties between work at the lab, family and the world."* His parents chose such a way of living in both Pasadena and Cold Spring Harbor.

Bohr, who blurred the lines between work, family, and the outside, also influenced Max's choice of lifestyle. Max remembered how important Bohr had been for him, both personally and professionally, and he wanted to be able, in his own way, to give to others what had been given to him. The Delbrücks did not have a simple country house like the one near Copenhagen that the Bohrs and physicists retreated to on weekends, but they replaced it with communal camping trips into the desert near Los Angeles and with summers in Cold Spring Harbor, times when young and old would mix. Along the way, Manny and Max energized, entertained, and educated a new generation of scientists.

27   The New Manchester

I believe that Caltech in the coming years will be to biology what
Manchester was to physics in the 1910s.

Max, writing to Niels Bohr in 1947

George Beadle, the man who recruited Max to Caltech, was
only three years older than Max. Born on a farm in Wahoo, Nebraska,
he probably would have become a farmer if not for a high school
teacher who encouraged him to attend the University of Nebraska.
In later life, even when he was a famous Nobel Prize–winning geneticist, Beadle always kept a little garden where he could grow a few
vegetables. He studied agronomy as an undergraduate and then went
to Cornell University, writing a Ph.D. dissertation on the genetics of
corn. Afterward, winning a fellowship that took him to Caltech, he
switched to cellular biology and worked on *Drosophila* in Morgan's
Fly Room. Following that, first with Boris Ephrussi and then with
Edward Tatum, he pioneered combining genetics with biochemistry, setting an example for the kind of disciplinary flexibility he was
beginning to advocate.

When Beadle was invited in 1945 to deliver the Harvey Lecture at
the Academy of Medicine in New York, he chose as his topic "The
Genetic Control of Biochemical Reactions." He had been feeling frustrated by academic barriers blocking the development of what he
saw as the future of genetics, and he planned to use this occasion to

expound his views. During the lecture he railed against *"the inflexible organization of our institutions of higher learning. The gene does not recognize the distinction—we should at least minimize it."* At the lecture's conclusion, Beadle said that in the past, university students had entered either a door marked "Genetics Laboratory" or another marked "Biochemistry Laboratory." He now proclaimed that in the future there would be only one door.

Linus Pauling was preaching the same message. Like Beadle, Pauling had gone to an agricultural school in his home state, in his case Oregon. He moved down the West Coast in 1925 for graduate studies at Caltech, a school then little older than a decade, and remained there, joining the faculty and becoming one of its brightest stars. By the mid-1940s he was recognized as one of the greatest and most influential theoretical chemists, perhaps the world's foremost expert on chemical bonding. Pauling was also a man with strong opinions, never one to back away from an issue he felt passionately about. In later years this persistence led to his being awarded the Nobel Peace Prize, to be added to the one he had already received in chemistry.

Pauling had also for years been exploring biological applications of chemical structures and, like Beadle, he was now looking for a united approach to chemistry and biology. At Pauling's urging, Beadle had been offered the position of chair of Caltech's Biology Division, and Max was one of Beadle's first recruits. Beadle knew Max well from their time together in the 1930s at Caltech and, more recently, their summers at Cold Spring Harbor. Max and Pauling were also friends and respected each other's work.

Max received the surprise offer of a professorship at Caltech in mid-December 1946, only six weeks after the George Washington University conference he and Beadle had attended. He accepted on December 27, writing Bohr several days later about how happy he was with his decision, *"because I believe that Caltech in the coming years will be to biology what Manchester was to physics in the 1910s."* Knowing that Bohr had begun to lay the groundwork for his revolutionary theory of the hydrogen atom in Manchester, Max was sure his old mentor would take notice of what he was saying. But what did Max think would be happening in biology? What in this new field would be the analog of Rutherford's experiments showing that the atom has a nucleus? Six years later, after Jim Watson wrote Max from Cambridge telling him about his and Crick's uncovering of the structure of DNA, Max sent Bohr another letter, telling him that he now knew the answer to that question. But in 1946 he was just dreaming of what might be discovered. As Hershey aptly described him, Max *"kept his eye on the big questions, even before they could be put into words."*

In his January 1947 letter to Bohr, Max also says that the move to Caltech *"signals the completion of my metamorphosis into a biologist."* Afterward, he never seemed to be quite so sure that his metamorphosis was complete. Whether as a physicist or a biologist, however, he was certainly excited by the prospect of starting research and teaching at Caltech, having stimulating colleagues, and working together with very bright graduate and undergraduate students. He also knew that both Beadle and Pauling, who, like him, had moved from field to field, would make sure no departmental or divisional barriers were thrown up. Interesting science, not protection of turf, would be the hallmark at their institution.

In 1948 there were still vacant lots in Pasadena, the small city near Los Angeles where Caltech is located. Manny and Max bought one and built a house there. It was only a few blocks from the campus, allowing Max to walk to and from work and to return to the lab in the evening to check on experiments. The home's furnishings were elegant but spartan, with no radio, television, or dishwasher; Max believed washing up after dinner should be a communal affair. An iconoclastic mix of decorations filled the house, but Max and Manny were an unconventional pair, and the furnishings suited them. The combined large living, dining, and kitchen area eventually held a grand piano, Indian rugs purchased in the 1950s on summer camping trips in Arizona, and several paintings by Jeanne Mammen, an artist Max had befriended in Berlin during the 1930s. Sixteen years older than Max, she had been a popular magazine illustrator as well as a painter at a time when social commentary was a prominent feature of pictorial art. That style was not tolerated under Hitler, and out of necessity, Mammen moved increasingly toward abstraction. Throughout all these ups and downs, her friendship with Max, and later with Manny, remained very close.

Max was conscious and even proud of his ancestry, but since he also seemed to feel the need to keep his distance from it, his connections to the past were often marked by irony. As an example, though he made an effort after World War II to assemble his father's collection of Goethe's works and to restore the volumes, the carefully bound editions were then shelved in a special bookcase in the bathroom, not his study. Also, in contrast to the Berlin tradition he grew up with, formality at work was anathema, starting with his own person: he was simply "Max" to everybody. He detested pomposity and delighted in poking fun at pretentious people in his speech, manners, and dress. Visitors were sometimes shocked. When an English scientist planning to accompany Max and Manny to a play appeared at their door, he was greeted by Max dressed as a pregnant woman and Manny as the "woman's" English husband, complete with bowler, mustache, and furled umbrella.

The Delbrück Pasadena house was a kind of second home for graduate students and postdoctoral fellows; in many ways they became part of the extended family, often with Manny looking after them. There were also the legendary Delbrück camping trips into the desert, simple expeditions organized on the spot, with plain food cooked on a campfire, some ordinary wine, hikes, a little scrambling over the rocks, and wide-ranging discussions about almost everything. Manny would arrange the excursions, pick Max up at the lab in their old car, and sometimes drive it into the desert until its wheels became stuck in the sand. That's where they would camp. In the evening you might find the Delbrücks, students, postdocs, and distinguished visitors from

America and abroad all huddled together by the campfire. Afterward, sleeping bags were brought out, simple food was cooked over a grill, and, as Jonathan Delbrück remembered, *"At night there was talk of all manner of things, as we clustered for dinner around the campfire, like a primeval tribal family, with Manny at its core and Max at the helm."*

These camping trips, part of the Delbrück mystique, were often punctuated by Max's practical jokes, which varied from his slipping rocks on the sly into people's knapsacks to hiding surreptitiously behind boulders on hikes into canyons after telling trailing hikers to follow him. Almost everybody, with the notable exception of Luria, loved the excursions, and foreign visitors would talk about them for years. The nighttime stars in those clear skies also enabled Max to return to a habit of his adolescence in Grunewald: careful nighttime sighting and recording of the stars' motion. This, too, became a remembered feature of those trips.

The Delbrücks' lifestyle was indeed a California version of the Bohrs' in Copenhagen, with their house becoming a California variant of the apartment directly above the Institute for Theoretical Physics, and the later Carlsberg mansion in which the Bohr family lived, while the Bohr Tisvilde country house was replaced by the Joshua Tree National Park's high desert. More remotely, the discussions around the campfire might even have reminded Max of the Sunday evenings in Grunewald when the Delbrück, von Harnack, and Bonhoeffer families got together.

28 Alpha, Beta, Gamma

The alpha, beta, gamma paper marked the birth of the hot Big Bang cosmology and started the march to precision cosmology. It is also exhibit one in my case that an interestingly wrong paper can be far more important than a trivially right paper.

Michael Turner, on a paper by Alpher, Bethe, and Gamow

The changes in Max's personal life after the 1946 conference on "The Physics of Living Matter"—the birth of his son, Jonathan, and, in the summer of 1948, of his daughter, Nicola; the visit to Germany; and the move to Pasadena—went hand in hand with changes in his professional life. The growing excitement in bacterial genetics and the study of bacteriophage meant that Max's career was reaching its apex and Caltech was coming to be seen as one of the world's great centers in research on the subject, much as Copenhagen had been for quantum mechanics. The influence of Cold Spring Harbor summers was also growing, and Max's position as Pope of the Phage Church was unquestioned.

By contrast, little had changed outwardly in Geo's life during this two-year period. Unlike Max, who felt the need to go back to his homeland as soon after the war as he could, Geo had no interest in returning to Russia. Perhaps if the thaw we know as glasnost had come a quarter century earlier, things might have been different. He certainly loved Russia—not what it had become, but the Mother Russia of his memory, the one he saw in his mind when reciting the reams of Russian poetry he had memorized. He feared, however, the new Russia, the one he had written about to Bohr in October 1945, the one that for him represented *"a great Deluge coming from the East which is bound to engulf the free man on the Earth."* And he had reason for these worries. As Churchill said in a famous speech delivered only months after Geo wrote that letter to Bohr, *"from Stettin in the Baltic to Trieste in the Adriatic an iron curtain has descended across the Continent."*

Geo masked his apprehensions under an appearance of lightheartedness, but they were never far from the surface. His nervousness was heightened on trips outside the United States, where he was haunted by the possibility of being kidnapped by Soviet agents, brought back to Russia, tried, and either executed or confined to a gulag. The fears never paralyzed him, but he remained wary, refusing for instance to stop in Hong Kong on a trip to India in the late 1950s because the city was adjacent to Red China.

Geo's thinking about science in 1948 was still focused on the subject of his landmark 1946 paper "Expanding Universe and the Origin of Elements." It had been known at the time that stars were roughly 99 percent hydrogen and helium, so the question of where and how the other elements of the periodic table were formed had been a natural one to ask. Geo conjectured that their assemblage occurred in a primordial, very dense expanding universe like the one Friedmann and Lemaître had proposed. The key new idea he now introduced was the need to consider nucleus formation as the universe rapidly expanded.

Geo imagined beginning his scenario with a dense gas of neutrons. As time passed, these neutrons would begin to decay into protons and electrons, and larger and larger nuclei would form by successive attachments of protons and neutrons. Eventually the still-expanding and still-cooling universe would have reached the point where electrons connected to nuclei. Atoms would then come into existence.

Much of this had been pure speculation, but two things happened in 1947 that fleshed out the skeleton. The first was that a very able graduate student, Ralph Alpher, joined Geo in analyzing the problem; the second was that experimental data on the capture of neutrons by heavier nuclei became available. With this new information, numerical estimates of element abundance became feasible. In his 1946 paper, Geo had simply assumed that neutrons *"coagulated"* together; now there were some data to use as input for how this might occur.

Nuclear formation in Alpher and Gamow's model began when the universe's temperature had cooled to the point where a neutron and a proton could bind together to form a deuterium nucleus—this point was approximately a billion degrees Celsius. From then on, heavier nucleus formation was assumed to proceed apace in a cooling, expanding universe. Preliminary results were suggestive, hinting at a possible explanation of all nuclear abundances.

By February 1948 Geo and Alpher were ready to submit a short paper for publication in *Physical Review.* Appearing on April 1, 1948, the paper contains a classic Gamow joke. Although Alpher and Gamow did the work, Geo could not resist an alliterative reference to the Greek alphabet; in submitting the paper, he added his friend Hans Bethe as middle author without informing him. The journal, unaware of what Geo was doing, simply changed Bethe's institutional affiliation from "in absentia" to Cornell University. The editors quite naturally assumed that Bethe had contributed to the article and the mistaken institutional affiliation was simply an oversight. This is how the famous "Alpha, Beta, Gamma" paper, as it came to be known, originated. When Bethe found out about the publication, his response was relaxed. He later told Alpher, *"I felt at the time that it was a rather nice joke and that the paper had a chance to be correct, so that I did not mind my name being added."*

Geo invited Hans Bethe to be a member of the examining committee when Alpher defended his Ph.D. thesis in 1948, so that "Alpha, Beta, and Gamma" could be united for the occasion. Not surprisingly, there were no problems, and Alpha became Dr. Alpha. Most such defenses—and I have sat through more than I care to remember—are attended by only a few friends in addition to the four or five faculty examiners, but Alpher's drew a huge crowd because news of it had been mailed out to George Washington University alumni and was picked up first by *Science News*, then reached the nationwide press. The topic of the creation of elements in the early universe caught the public's fancy. One of the conclusions of the thesis—that the formation of nuclei was concluded in about three hundred seconds—even led to a famous cartoon by the *Washington Post's* Herblock of a frowning bomb-shaped man sitting in an armchair considering the announcement that the world was created in five minutes (see figure 2). Geo, who loved publicity, enjoyed the ensuing fanfare. But he and Alpher went quickly back to work.

In 1948, in order to help with the study of the early universe's

FIGURE 2

A 1948 Herblock Cartoon, copyright by The Herb Block Foundation.

workings, Alpher asked a friend to join the collaboration. While still a graduate student at George Washington University, he had taken a position at the Johns Hopkins Applied Physics Laboratory near Silver Spring, Maryland, and there met Bob Herman, a few years older than he was and already an experienced physicist. Herman had received a Ph.D. at Princeton and knew a great deal about the areas of physics that interested Alpher and Geo at the moment. Since he was the child of Russian immigrants, he also knew enough Russian to appreciate Geo's declaiming Pushkin's verse. Herman was happy to start working with Alpha and Gamma, but did draw the line at acceding to Geo's request that he change his last name to Delter. This did not, however, stop Geo from inserting in a later paper a reference to work by *"Alpher, Bethe, Gamow, and Delter."*

Alpher and Gamow, now working with Herman, realized they had previously neglected to include photons in the early universe's energy balance. This was a major omission in their discussion, because photons, the quanta of electromagnetic radiation, have energies that are proportional to temperature when conditions are like those envisioned when nuclei are being formed. Later, as the universe cools and expands, the energy balance shifts; the scales increasingly tilt toward matter and away from radiation as the energy of each photon diminishes. Alpher, Gamow, and Herman commented on what these photon energies might be expected to be in the present-day universe, but neither they nor anybody else at the time pursued the question of whether these photons could be detected. There were no insurmountable technical problems, but progress in science is often not as simple as it seems in hindsight. Fifteen years later, observers who were unaware of the predictions Alpher, Gamow, and Herman had made finally did detect the radiation.

Alpher, Gamow, and Herman's work created a big stir in 1948, but it was quickly criticized for not delivering what it had promised: a successful proposal for how to synthesize atomic nuclei. It seemed to explain why the universe is composed overwhelmingly of hydrogen and helium, but it stumbled in trying to show how all the other nuclei are produced. The reason for why it failed is simple. Starting out with a universe rich in neutrons and protons, it is relatively easy to form a helium nucleus (two neutrons and two protons), while a hydrogen nucleus (nothing but a single proton) is there from the start. But reaching nuclei past hydrogen or helium requires an intermediate state that is simply not present.

There are *no* stable nuclei having a total of either five or eight protons and neutrons, and without them, one cannot go beyond hydrogen and helium in the early universe. A proton or a neutron can collide with a helium nucleus, but the resulting combination $(1+4=5)$ will break up before providing a stepping-stone to larger nuclei; the same is true for an encounter between two helium nuclei $(4+4=8)$. The former combination requires a stable nucleus with five protons and neutrons, and the latter, one with eight. Neither exists. Nor can one imagine three nuclei coming together at once, too unlikely an event in a rapidly expanding universe. In plain language, it seemed easy to create hydrogen and helium, but impossible to go beyond that.

The predicament was tantalizing. Many fine minds besides those of the original three pioneers who proposed the model for an early universe looked for ways to bridge the gap, but nobody succeeded. The general opinion in the physics community that followed these efforts was that although the theory was stimulating, it was wrong.

But as Michael Turner, a distinguished University of Chicago astrophysicist, has said, it was *"interestingly wrong."* In 2008 he wrote an article in *Physics Today* about the "Alpha, Beta, Gamma" paper.

You'd have to be living in a cave in Afghanistan not to know that cosmology is in the midst of an extraordinary period of discovery—perhaps even a golden age. But you might not know that it all started on April Fool's Day 60 years ago. Ralph Alpher, Hans Bethe and George Gamow published a Letter to the Editor entitled "The Origin of Chemical Elements" in the April 1 issue of the Physical Review. . . . *The alpha, beta, gamma paper marked the birth of the hot Big Bang cosmology and started the march to precision cosmology. It is also exhibit one in my case that an interestingly wrong paper can be far more important than a trivially right paper; recall Wolfgang Pauli's famous putdown, "It isn't even wrong."*

But what makes a paper interestingly wrong or trivially right? It should be said at the outset that we are discussing theoretical proposals. An experiment is right or wrong, never *"interestingly wrong,"* though I do admit the interpretation of an experimental result may be interestingly faulty. But that certainly isn't the same as getting a wrong answer. The distinction between interestingly wrong and trivially right theoretical proposals is far subtler. A trivially right paper may represent years of hard work and may well be a worthwhile contribution to a field, but it introduces no new ideas or techniques. A good deal of skill might have been involved in carrying out the work, but the answers are likely to be what one expected at the outset. An interestingly wrong paper, on the other hand, features the new and unexpected, points out unforeseen paths and suggestive insights. It is also almost never entirely wrong; it may overreach in its claims or need to be revised, extended, and reinterpreted, but it has brought something new into the world.

The classic twentieth-century-physics *"interestingly wrong"* paper is Bohr's 1913 theory of the atom. The picture he proposed of electrons circling the atomic nucleus on circular orbits whose radii are fixed by quantum rules is wrong. And yet it explained with extraordinary precision many of the features of atoms, so much so that in the ensuing decade, the central problem in quantum physics became that of finding a theory that incorporated the desirable attributes of Bohr's theory while shedding the ones that contradicted experimental results. Quantum mechanics can be traced back to that *"interestingly wrong"* paper.

Though not rising to the overall significance of Bohr's 1913 work, the contribution of "Alpha, Beta, and Gamma" is deservedly remembered. The paper was wrong in thinking that this was the way the periodic table's elements were formed, but it did provide an important step forward by displaying the correct and very interesting idea that the universe originated in a sudden explosion, followed by a rapid cooling as space expanded.

29   Big Bang Versus Steady State

It is true that Gamow was funny and that he drank. It is also true that
he was a brilliant scientist, devoted friend and concerned teacher,
whose intuition exceeded that of any scientist I have known.

Vera Rubin, remembering Geo

While Geo and his co-workers were encountering difficulties with their picture of the universe's evolution, a competing theory was adding to their woes by presenting a counterproposal that
seemed more satisfactory to many on both philosophical and scientific grounds. Whereas Geo's model was predicated on an expanding
universe that had begun in an explosion, the rival model proclaimed
there had been no beginning and would be no end. Three Cambridge
scientists, Hermann Bondi, Thomas Gold, and Fred Hoyle, respectively a mathematician, an engineer, and a physicist, produced the
contender. Bondi and Gold, both Austrian Jewish refugees, had met
in 1940 when they were deported from England and interned with
other notable young refugee scientists in a prisoner of war camp outside of Quebec. After little more than a year, when it was finally realized that these internees were eager to contribute to the fight against
Nazi Germany, Bondi and Gold were returned to England and, shortly
thereafter, sent to work on radar systems. Fred Hoyle was the head
of the unit to which they were attached; the three soon became close

friends. Discussing in their spare time all kinds of scientific problems, they found their interests converging on questions of cosmology.

These discussions led during 1947 to their formulation of the steady state theory of the universe. Though much of the effort was common, there were sufficiently many differences between the approaches of Bondi and Gold, on the one hand, and of Hoyle, on the other, to warrant publishing separately. However, the essential common feature of both versions was that the density of matter in the universe, when viewed on a large scale, remained unchanged. This seemed to run counter to the recognized observation that the universe was expanding, so they postulated that matter was being continuously created at the rate necessary to balance the dilution caused by the universe's spreading out. A quick calculation showed them that the desired production was small enough to have evaded detection by known experiments. While Hoyle accommodated this creation by modifying Einstein's equations of general relativity, Bondi and Gold did not commit themselves to a specific solution. In both cases the universe was seen as having no beginning and no end.

The upshot was that in 1948 there were two very different proposals for the universe's evolution, one saying that it was eternal and the other asserting that it had begun in a cosmic explosion. Furthermore, two very dynamic advocates, each of whom had a penchant for arguing his case in public, made presentations of the two very different points of view. Fred Hoyle took the lead in a set of lectures he delivered on the BBC during the spring of 1949. These were quickly converted into a book, *The Nature of the Universe,* which became a bestseller. Hoyle, who later also became well known as a science fiction writer, was a gifted author, with a good ear for phrases and a combative streak in his writing and speaking. During his lectures and in the book he derisively referred to Geo's point of view about the universe's origin as the "Big Bang." Not surprisingly, he found it far inferior to his own steady state theory. In an ironic twist, Hoyle unwittingly helped popularize his opponents' theory by giving it such a memorable name.

Nor did the pugnacious Hoyle limit his attacks to the Big Bang. He concluded his book with an assault on Christianity, wanting to make sure that the idea of a steady state universe would not be used as support for the notion of an immortal soul, a concept warranting even more scorn from him than Big Bang cosmology.

Geo rose to the challenge in the fall of 1952 with his own book, *The Creation of the Universe*. He placed the first bars of Haydn's *Creation* on the title page and dedicated the book *"to fellow cosmogonists of all lands and ages."* This group turned out to include one unlikely enthusiast: Eugenio Pacelli, or Pope Pius XII. Addressing the Pontifical Academy of Sciences shortly after the book's appearance, the Pope seemingly took sides in the debate by using the idea of a universe with a sudden beginning as evidence for the existence of a Creator. He described the cosmic beginning as a *"primordial Fiat Lux, when, along with matter, there burst forth from nothing a sea of light and radiation and the elements split and churned and formed into millions of galaxies."* The Pope's apparent choosing of a winner in a scientific debate enraged Hoyle and dismayed the devout priest/cosmologist Georges Lemaître, a member of the Pontifical Academy of Sciences. Despite being an early formulator of the notion of an expanding universe, Lemaître had always been very careful to keep science and theology separate and did not want to see that line crossed, most particularly not by his Spiritual Father.

But Gamow had no such qualms. Always on the lookout for a good joke, he prefaced his next cosmology paper, which appeared in February 1952, with a lengthy quote from the Pope's address, preceding it with a tongue-in-cheek remark: *"It can be considered now as an unquestionable truth that . . ."* Reading in the *New York Times* on May 24 that Sir Harold Spencer Jones, the Astronomer Royal of Great Britain, had felt the need to respond to the Pope's proclamation with an endorsement of the steady state model further increased his amusement with the whole situation. However, though Geo was finding this all deliciously comical, he now began to worry that he might be overdoing the religious angle, so he prefaced the second printing of his *Creation of the Universe* with a corrective note:

In view of the objections raised by some reviewers concerning the use of the word "creation" it should be explained that the author understands this term not in the sense of "making something out of nothing" but rather as "making something shapely out of shapelessness" as for example in the phrase "the latest creation of Parisian fashion."

But beyond all this semicomical sparring, a serious scientific debate was going on. By *"shapely out of shapelessness,"* Geo was maintaining that all nuclei were constructed in the early universe, using as building blocks the protons and neutrons that had been there almost from the beginning. Though discouraged by the difficulties of synthesizing the nuclei of elements other than hydrogen and helium, he was far from ready to surrender. Moreover, he was still making fun of those, like Hoyle, who were willing *"to assume that different atoms were cooked at different places and under different temperature and pressure conditions."* Driving home his point in plain language, Geo compared these individuals to the *"inexperienced housewife who wanted three electric ovens for cooking a dinner: one for the turkey, one for the potatoes and one for the pie."*

But despite Geo's dismissal of those who called for many ovens, it was becoming increasingly clear that more than one would be needed. Ironically it was his brilliant opponent, Fred Hoyle, who spotted a second oven. He had begun as early as 1946 studying how heavier nuclei first appeared and had identified supernovae as the likely site where nuclei ranging from carbon to nickel were formed, going on to show how this could occur. Further evidence for creation in stellar cores came in 1951, when a young theoretical physicist, Edwin Salpeter, found that it was possible to fuse three helium nuclei under the temperature and pressure conditions operating in such an environment;

a carbon nucleus was the outcome. The nuclear gap that Alpher, Gamow, and Herman had been unable to bridge in their examination of the expanding early universe could be crossed in this second site. But even that wasn't the end of the story.

Two years later, in 1953, while he was spending a few months at Caltech, Fred Hoyle returned to his earlier studies and now found a flaw in Salpeter's scheme. The calculation of three helium nuclei forming a carbon nucleus was correct, but the picture was still in trouble. The reason was that the subsequent helium-plus-carbon reaction was so efficient that all of the carbon would be converted into oxygen. But carbon is known to be abundant.

There was a way out, but it required that an excited state of the carbon nucleus be produced, and such a state had not been observed. It was as if a stepping-stone needed to be found in order to cross a stream. Was there no stone, or had it simply not yet been seen? Hoyle consulted Willy Fowler, Caltech's expert experimentalist on nuclear reactions, asking him if the state's existence could have escaped detection in earlier scans. Fowler not only said that it could have but, realizing the question's importance, set to work immediately to see if the excited state existed.

In his autobiography, *Home Is Where the Wind Blows*, Hoyle describes very nicely the suspense a scientist feels as he waits to see if a theoretical prediction is confirmed. He conveys this by comparing the suspense he suffered during those two weeks of 1953 to that which a prisoner experiences waiting for the jury to reach its verdict:

> *Except, of course, the prisoner in dock knows already whether he is guilty or not. In court the prisoner hopes the jury gets it right if he knows he's innocent and he hopes the jury gets it wrong if he knows he's guilty. In physics, on the other hand, the jury of experimentalists can be taken always to be right. The problem is that you don't know whether you are innocent or guilty, which is what you stand there waiting to learn as the foreman of the jury gets up to speak.*

The jury's verdict came in: Hoyle had been right. His sensational prediction may be the only one both made and verified by the anthropic principle, simplistically stated as the notion that the universe is the way it is, for otherwise we would not exist to study it. According to Hoyle, if that excited carbon state had not been present, no carbon would have survived the chain of nuclear production. In its absence, there would have been no organic molecules, and hence no life. But life exists, and therefore so must the carbon state.

Geo was willing to give Hoyle credit for his prediction and admit that his own scenario of the formation of all the elements was not correct, but he certainly did not think this meant his overall point of view was wrong. He still held the belief that the universe had begun in what Hoyle had derisively called a Big Bang, an instant of infinite temperature and density. He further believed that a subsequent cooling and expansion during which nuclei at least began to form followed this instant. Geo was right, and his is the view we have now seen proved in exquisite detail.

Geo summarized his thoughts on astrophysics in a set of lectures he delivered at the University of Michigan in July 1953. The director of the university's observatory had obtained funds to bring together advanced graduate students and postdoctoral researchers for several weeks of intensive astrophysics discussions, to be led by two principal lecturers, Geo and Caltech's Walter Baade. Many of the future leaders of the field, altogether some fifty individuals, were in attendance.

Almost forty-five years later, Ed Salpeter, by now a celebrated elder astrophysicist, recalled how crucial those summer 1953 lectures were for his education. Thinking back on that period, he remembered Geo's *"humanity and exuberance"* as well as his *"interest in and speculation on amazingly different topics."* Geo seemed to be thinking far ahead of everybody else, with all-encompassing curiosity that ranged over the whole field of astrophysics. Ideas, many just glimmerings of possibilities, came gushing out.

Vera Rubin, another famous scientist who attended the Michigan lectures, echoes Salpeter. She remembers how Geo:

delivered brilliant lectures at a now legendary summer school at the University of Michigan that I attended as his Ph.D. student. It is true that Gamow was funny and that he drank. It is also true that he was a brilliant scientist, devoted friend and concerned teacher, whose intuition exceeded that of any scientist I have known.

I recently visited Dr. Rubin, one of the United States' most distinguished astronomers, at her office in the Carnegie Institution's Department of Terrestrial Magnetism, a beautiful set of buildings on a hill overlooking Rock Creek Park in Washington, D.C. I asked her to tell me more what she meant by her comment that Geo's *"intuition exceeded that of any scientist I have known."* She explained that she was alluding to Geo's special ability to think in ways nobody else did, to ask prescient questions that nobody else had thought of, and to let his imagination guide him into uncharted waters.

Her remark about Geo's drinking is echoed by others. It was widely known that by the 1950s he was drinking heavily. As far as his friends noted, this did not seem to affect his thinking abilities, but it did eventually cause him serious health problems.

As a graduate student, Rubin wrote her Ph.D. dissertation under Geo's supervision. Though he provided her with some ideas and direction, she readily admitted that he was less than diligent in providing help with the calculations. While she was writing it, he was away in California on sabbatical leave. When she sent it to him for inspection, he wrote back saying the dissertation looked *"very nice,"* but that he hadn't been able to bring himself to verify its contents. The truth is that dissertation advisers often do not check the details of their advisees' work, but only Geo would happily admit he hadn't. But then, as I have said before, he wasn't quite like anybody else.

My getting to know him during the summer of 1948 at the Cold
Spring Harbor Laboratory was a revelation, with his tall youthful
esprit soon becoming the model for what I wanted out of my own life.

James Watson, on Max

By 1948 Max's accomplishments at Caltech and at Cold Spring
Harbor were making him an iconic figure in molecular biology. In five
years he had gone from being a relatively obscure instructor at Vanderbilt to having a commanding position in an emerging field of science. He
was also seen as possessing such charisma that many admirers began to
feel about him the way the young James Watson did in describing him
as *"the model for what I wanted out of my own life."* Nobody was saying
this about Geo, and yet, with his extraordinary intuition, insight, and
imagination, there was much to admire. Comparing Geo and Max, I am
reminded of the ancient Greek adage *"The fox knows many things, but
the hedgehog knows one big truth."* Though Max's physical appearance
was more like a fox's and Geo's heftier bearing more like that of a hedgehog, it was Geo who focused on *"many things"* while Max continued
to pursue the question that had drawn him to biology a decade earlier:
How is genetic information replicated and transmitted?

By the end of the 1940s, while Geo's proposal for explaining the
origin of chemical elements was meeting seemingly insurmountable
obstacles, Max's search for his *"one big truth"* was advancing steadily.

A series of groundbreaking experiments in 1946 showed that bacteria had the capacity to exchange genetic information, and phages also seemed able to perform this task, but what was it they were exchanging? Until the 1950s, the prevailing belief was that genes were proteins, large complicated molecules composed of smaller subunits, amino acids, linked end to end. Since there were thought to be at least twenty different amino acids, a protein could be any of the combinations constructed with this large chemical alphabet, exactly what was needed to encode the genetic information required by life's innumerable variations. And yet there was evidence that genes were not protein molecules. Why had those clues been missed?

The paths to discovery are frequently veiled, and progress in science all too often resembles wandering through a maze rather than racing down a straightaway. Blind alleys meet you at every turn along the way, and the most interesting findings are often the unexpected ones. The path taken is seen when the race is over, and only then does the course appear obvious. The answer to what is a gene provides a lovely example. With the benefit of hindsight, the first clear indication that genes are segments of DNA rather than proteins came in 1944, but it still took another ten years before the notion gained credence.

Deoxyribonucleic acid, usually known by its acronym, DNA, is the key to genes' distinctiveness. This chemical was discovered in 1869 by a twenty-five-year-old Swiss biochemist named Friedrich Miescher while he was examining the contents of cell nuclei he had extracted from pus-soaked bandages discarded by a local hospital. He found that an interesting substance he called nuclein was located in the chromosomes, already a hint of its importance for genetics. Soon thereafter Miescher, by now a professor in his native Basel, set about studying nuclein's chemical structure, employing a more attractive source: sperm from salmon caught in the Rhine, the local river.

Working on the structure of nuclein together with students, Miescher isolated what we today call deoxyribonucleic acid. Further research established that there was also a second kind of nucleic acid.

Having a slightly different sugar—ribose instead of deoxyribose—it was given the name ribonucleic acid, or RNA.

Little other than the appearance of nucleic acids in cell nuclei warranted looking for a connection between the acids and heredity. But in 1928, Fred Griffith, a British public health microbiologist, made a suggestive discovery. Griffith was interested in bacterial pneumonia, a common and extremely dangerous disease in those pre-antibiotic days. The affliction is characterized by two common microbial strains, labeled R and S because of their respective "rough" and "smooth" appearance under a microscope. When injected into mice, R is harmless while S is lethal, a difference due to R bacteria lacking the protective capsule that encloses S bacteria. This covering acts as a shield, protecting the S bacteria from the antibodies mice produce to defend themselves. The capsule is, incidentally, also the reason for the smooth appearance under the microscope of S bacteria. Griffith found, as expected, that S bacteria could be rendered harmless by subjecting them to high temperatures, known to be lethal for bacteria.

Now came the shocking surprise. When he injected mice with the harmless R strain and the equally harmless heat-treated S strain, the mice died. How could two harmless kinds of bacteria kill a mouse? Something miraculous had happened. When dissected, the mice were seen to be full of live S bacteria. One possibility was that a genetic mutation had occurred in the R bacteria, allowing them to grow a protective capsule, but Griffith never considered this prospect, probably because bacteria were not thought to have such a capability. He believed instead that R bacteria had always possessed the ability to grow capsules but had been unable to do so until they had been provided with the proper nourishment, perhaps given to them by the dead S bacteria.

Griffith was a shy, retiring man, so the news of his result went relatively unnoticed, though a few scientists followed it up, even achieving in a test tube the same pneumonia bacteria transformation that had previously been seen in mice. This aroused the interest of Oswald Avery. He was a doctor who had joined the staff of the Rockefeller Institute in 1913 and would remain there for the rest of his career. Avery, called by many simply The Professor, was the prototype of the dedicated scientist, enormously hardworking, modest, and self-effacing. He dressed meticulously, and wore a pince-nez to focus his vision. Beloved by students and colleagues, he was a bachelor with few interests outside his research.

Over the course of almost a decade, from the mid-1930s to the mid-1940s, Avery performed a series of experiments, many of them with his colleagues Colin MacLeod and Maclyn McCarty, that led to an exquisitely detailed understanding of how the information about growing a capsule was transmitted from the heat-treated S bacteria to the R bacteria. The so-called transforming factor, the agent that carried the instructions from one bacterium to the other, remained unaffected by reactions that stripped the S bacteria of their outer shell or by ones that removed their proteins. At the end, only two candidates were left: RNA and DNA. An enzyme that cleaved RNA did not affect the result, leaving only DNA as a possible transforming factor. As a check, Avery performed the experiment in reverse, first introducing an enzyme that cleaved DNA but left RNA and proteins intact. When this was done, the R bacteria no longer grew capsules. Without a doubt, the transforming factor, the carrier of the genetic message, was DNA.

You might think that Max, Luria, and others intent on discovering the identity of genes ignored or misunderstood Avery's work because their work seemed to have been unaffected by his findings. Max was even privy to Avery's private thoughts about the transforming principle, as Avery's brother Roy was a professor of microbiology at Vanderbilt's Medical School and hence a colleague of Max's at the time. The younger Avery received frequent letters from his brother. In May 1943 he showed Max a seventeen-page letter that had just arrived. Written six months before the key paper by Avery was submitted, the letter is a remarkable document, clearly showing its author's excitement in a way that is almost always removed from a journal article:

> *If we are right, and of course that is not yet proven, then it means that nucleic acids are not merely structurally important but functionally active substances in determining the biochemical activities and specific characteristics of cells and that by means of a known chemical substance it is possible to induce predictable and hereditary changes in cells. This is something that has long been the dream of geneticists. . . .*
>
> *Sounds like a virus—may be a gene. But with mechanisms I am not now concerned. One step at a time and the first step is what is the chemical nature of the transforming principle. Someone else can work out the rest. Of course the problem bristles with implications.*

Toward the end of the letter, the cautious and critical Avery reminds himself that *"It is a lot of fun to blow bubbles but it is wiser to prick them yourself before someone else tries to,"* an attitude that is certainly very different from the one Geo held. He enjoyed *"blowing bubbles"* as much as anyone but was also perfectly happy to see others prick them. That just meant he had to blow more bubbles.

Avery died in 1955, after the importance of DNA had been universally recognized but before he could be awarded the Nobel Prize he so richly deserved. Years later, preparing for a symposium honoring Avery's discovery, Max asked Roy Avery to look for the letter he had received from his brother more than a decade earlier. After several days of searching, Roy found it, and Max read the key parts to a fascinated audience, opening for them a window on science in the making.

Avery's natural reticence was a contributing factor to his discovery's not being seen as the smoking gun indicating that DNA carries genetic information. When the definitive paper appeared in a 1944 volume of the *Journal of Experimental Medicine*, Avery was sixty-seven years old, nearing the end of a career marked by careful laboratory work but without extraordinary results. Furthermore, though he might have revealed in private to his brother his dream of DNA's importance, he understated his conclusions when presenting them to the public. The crucial 1944 paper by Avery, MacLeod, and McCarty asserts only:

> *If the results of the present study on the chemical nature of the transforming principle are confirmed, then nucleic acids must be regarded as possessing biological specificity the chemical basis of which is as yet undetermined.*

However, Avery's reserve wasn't the main reason DNA was not thought of as the principal carrier of genetic information. Beadle and Tatum's "one gene, one enzyme" model had made researchers believe that genes were proteins. Even more important, the nucleic acids' chemical structure seemed far too simple to code for life's complexities.

DNA and RNA have three ingredients: two of them, a sugar and a phosphate, repeat themselves over and over again in the assembly of the large molecule. The third, commonly known as a base, is the key. By early in the twentieth century, the nucleic acids' five bases had already been identified: they are guanine (G), adenine (A), cytosine (C), thymine (T), and uracil (U), names with rather prosaic origins— for example, thymine was first found in thymus glands. Both nucleic acids were known to have G, C, and A bases, but U replaced T in RNA.

The majority of chemists in the mid-1940s still thought that the order of A, G, C, and T bases in DNA was boringly repetitive, meaning it could not possibly convey the complicated message necessary for genetic transmission, but this notion changed toward the end of the decade. The Columbia University biochemist Erwin Chargaff then showed that the amounts of G and C are always equal, as are those of A and T, but the relative proportion of the two pairs differs from one organism's DNA to another's. Chargaff's rules, as they were known, made it seem at least possible that DNA was genetic material because the variation of ratios could conceivably convey genetic information.

Furthermore, a French group showed that sperm cell nuclei had only half as much DNA as other cells in an organism, which was consistent with the notion that an offspring should receive half its blueprint from the father and half from the mother. All this was suggestive, but not enough to convince doubters.

In 1952 Alfred Hershey, who with Max and Luria formed the Phage Church's trinity, clinched the argument that DNA was the genetic information carrier. Working with Martha Chase, Hershey conducted the "Waring Blender Experiment," so known because of his and Chase's remarkable use of this simple kitchen appliance to separate bacteria from phage protein. In essence, their experiment showed what viruses (by then identified as DNA with a protein coating) did when they settled on bacterial cells. The virus's DNA was injected through the cell wall, with its protein coating left behind. Even though the coating was then shaken loose from the bacteria by the action of the blender, the bacteria were still altered; in other words, DNA and not protein accomplished the virus's takeover of the bacteria workings. But this still left open the question of how it performed its task.

31 The Double Helix

I have a feeling that if your structure is true and if its suggestions concerning the nature of replication have any validity at all, then all hell will break loose, and theoretical biology will enter into a most tumultuous phase.

Max, in a 1953 letter to James Watson

The answer was discovered in Cavendish Laboratory in Cambridge, England, in 1953. More than twenty years earlier, in 1929, Geo had spent a term as a Rockefeller fellow at the Cavendish, invited there by Rutherford at Bohr's urging. That was the heyday of the laboratory's reign in nuclear physics. But within a few years of Geo's stay, the heady period had come to an end, as cyclotrons proved more effective in exploring the frontiers of nuclear physics than the Cavendish Laboratory's ingenious and inexpensive tabletop experiments. When, in 1937, Rutherford died unexpectedly after a routine hernia operation, Sir Lawrence Bragg was chosen to succeed him as Cavendish head. A dominant figure in the use of X rays to determine molecular structure, Bragg had been awarded the Nobel Prize in Physics in 1915 when he was twenty-five years old, and he had not slowed down since then.

This was in some ways an unexpected selection, since it marked a radical departure from the laboratory's primary focus on nuclear physics, but Bragg had the intellect, the personality, and the credentials to warrant the selection. It is very rare for a research institute that has been extraordinarily successful in one line of research, in this case nuclear physics, to shift its emphasis, even if it does see its powers in the field declining. But the Cavendish under Bragg rejuvenated itself and was handsomely repaid for the transformation by an extraordinary set of achievements in completely different fields, one of which was the structure of living organisms.

By the late 1930s X rays had been used successfully to analyze the placement of atoms in simple crystals, but proteins, coiled molecules with thousands of atoms, represented a challenge of a different magnitude. And yet a true chemical approach to biology was not possible without this information, since proteins' biochemical action depends on the three-dimensional arrangement of the atoms. The flamboyant British scientist John Desmond Bernal, nicknamed Sage and, occasionally, Great Sage because he reportedly knew everything, was the leading advocate of this approach to biochemistry. So in 1936, when a twenty-two-year-old recent college graduate named Max Perutz went to consult him on a choice of career, Sage said to him, *"The secret of life lies in the structure of proteins and X-ray crystallography is the only way to solve it."* Taking Sage's advice to heart, Perutz began using the technique Sage proposed in order to study hemoglobin, the red blood cells' protein, which transports oxygen from the lungs to the body tissues where it is needed.

Four years later, in 1940, Perutz, by now a fresh Cambridge Ph.D., found himself together with Hermann Bondi and Tommy Gold, the future steady-state-universe theorists, on a boat bound for a prisoner of war camp in Canada. He, like the other two, was an Austrian Jew and was being sent away by the British government as a potentially dangerous enemy alien. Slightly older, he became the organizer of the science courses the three of them and a few others conducted in that camp. Released after a little over a year, when the intelligence units concluded the three were an asset and not a liability, they returned to England and began working on defense projects. The war's conclusion found all three back in Cambridge, ready to resume basic research. In Perutz's case, this meant returning to his study of hemoglobin.

Six years later, a curious visitor appeared in his laboratory. Perutz remembered the occasion:

> One day in September 1951 a strange young head with a crew-cut and bulging eyes popped through my door and asked, without saying as much as hello, "Can I come and work here?" He was Jim Watson, who wanted to join the small team of enthusiasts for molecular biology which I led at the Physics Laboratory in Cambridge, England.
>
> My colleagues were John Kendrew, a chemist like myself, and Francis Crick and Hugh Huxley, both physicists. We shared the belief that the nature of life could only be understood by getting to know the atomic structure of living matter and that physics and chemistry would open the door, if only we could find it.

Perutz invited the young American to join the "*small team*," and so Watson met Crick, a man whose inventive sense Watson describes on the second page of his famous memoir, *The Double Helix*, in words reminiscent of Teller's portrayal of Geo:

> Often he came up with something novel, would become enormously excited and would tell it to anyone who would listen. A day or so later he would often realize that his theory did not work and return to experiments, until boredom generated a new attack on theory.

Meanwhile, Perutz and his co-workers were continuing their meticulous, dedicated effort to decipher the structure of the enormously complicated hemoglobin model. It would take them almost another ten years to complete the task, but when it was over, the magnitude of the feat was immediately recognized. John Kendrew and Max Perutz received the Nobel Prize in Chemistry in 1962. They shared the stage in Stockholm with Crick and Watson, who had deciphered the structure of DNA only eighteen months after the *"strange young head"* popped through the door. Watson and Crick came to the problem of DNA's structure with complementary expertise; Crick was versed in the English crystallographic analysis school, and Watson was a phage group geneticist. This was the right mix of skills to crack the problem, though their brilliance, ambition, and drive should not be discounted.

Their Nobel Prize, in the field of Physiology or Medicine, was awarded for *"their discoveries concerning the molecular structure of nucleic acids and its significance for information transfer in living material."* The duo's success in constructing their double helix model was also due to their access to the beautiful X-ray pictures of DNA taken at King's College in London by Rosalind Franklin and Maurice Wilkins. Appropriately, Wilkins shared the Nobel Prize with Crick and Watson. Franklin, who obtained the crucial data on which the Watson-Crick model was based, had tragically died of cancer in 1958; she was only thirty-eight years old.

Watson and Crick had believed that determining the placement in DNA of phosphates, sugars, and above all the bases was the key to determining why this molecule seemed to play such a crucial role in genetics. Six thousand miles away, in Pasadena, Max's friend and colleague Linus Pauling had come to the same conclusion and was also trying to construct a model for DNA. Crick and Watson seemed to have little chance of outracing Pauling, the world's foremost expert on molecular binding. Making this even more unlikely, Pauling already had experience dealing with a similar problem: two years earlier he had pulled off a coup by showing how chains of amino acids arrange themselves in a helical shape. However, as we know, Watson and Crick unexpectedly won that race.

Pauling had thought that DNA was composed of three helices wound around one another, whereas Watson and Crick had decided, correctly, that it was a double helix. Furthermore, their model had an elegant feature: since base A on one chain would always bind to base T on the other, and base G always to base C, knowing the sequence of bases on one chain meant the other chain's sequence would be specified. In other words, each chain would carry complementary information. If they could be separated, each could serve as a template for further production.

Watson and Crick's first letter to *Nature* appeared on April 25, 1953. It concludes with the tantalizing phrase *"it has not escaped our notice that the specific pairing we have postulated immediately suggests a possible copying mechanism for the genetic material."* Though not spelled out, what they meant was clear: all genetic information was contained in the ordering of the bases along each of the two intertwined chains of the helix. This was the great breakthrough that changed genetics forever. Finding out what sequence corresponded to any given gene was still a long way off, but in principle, the subject had now entered a new and far less mysterious phase.

Max had been kept abreast of developments. Knowing Watson from Cold Spring Harbor, he had recognized his talent and offered him a fellowship at Caltech with funds from the Polio Foundation. It was scheduled to begin in September 1952, but when Watson, excited by his research on DNA, decided he wanted to stay an extra year in Cambridge, Max approved and quickly arranged for the funds to be transferred from Caltech to Cambridge. Watson was grateful. Continuing to look for Max's support, he wrote to him frequently, keeping him up to date on his progress.

This meant that Max was the first person in the United States to hear of the double helix model, since Watson sent him a long letter as soon as he and Crick had reached their conclusion. Dated March 12, 1953, it is appended to Watson's *The Double Helix*. I recently asked Watson why he had first written about DNA to Max instead of to Luria, who was, after all, his thesis adviser and in some sense his sponsor. The answer was quite simple: after Cambridge, he was planning to go to Caltech, so he was looking to the future rather than the past. He went on to add that he also felt that Max would immediately grasp the beauty of the double helix model and would appreciate its significance more than Luria would.

However, still afraid that something about their picture was wrong, Watson concluded the letter with a postscript: *"We would prefer your not mentioning this letter to Pauling. When our letter to* Nature *is completed we shall send him a copy."* Max ignored this request, since Pauling had asked him to be kept informed and because, as Watson points out, *"Delbrück hated any form of secrecy in scientific matters."* Watson need not have worried; the model was right.

Max had realized the paper's importance and was even more excited after talking to Pauling and a few other colleagues. Pauling conceded that Watson and Crick had won the competition to discover DNA's structure. Shortly after that, on April 14, 1953, Max wrote two letters. One, a short one to Bohr, concludes, *"Very remarkable things are happening in biology. I think that Jim Watson has made a discovery which may rival that of Rutherford in 1911,"* an unmistakable allusion to Rutherford's discovery of the atomic nucleus. The other letter, addressed to Watson, reads:

> *I have a feeling that if your structure is true and if its suggestions concerning the nature of replication have any validity at all, then all hell will break loose, and theoretical biology will enter into a most tumultuous phase.*

Peter Medawar, a Nobel Prize winner for deciphering how the immune system accepts or rejects transplants, wrote about biologists' reactions to this event in his review of Watson's *The Double Helix:*

> *The great thing about their discovery was its completeness, its air of finality. If Watson and Crick had been seen groping toward an answer; if they had published a partially right solution and had been obliged to follow it up with corrections and glosses, some of*

them made by other people; if the solution had come out piecemeal instead of in a blaze of understanding; then it would still have been a great episode in biological history but something more in the common run of things; something splendidly well done, but not done in the grand romantic manner.

The double helix model and its implications for genetics were a great surprise to Max. He had not shared the Cambridge group's or Pauling's belief that determining DNA's structure would lead to a revolution in genetics. As Crick writes in his autobiography:

I don't think Delbrück cared much for chemistry. Like most physicists, he regarded chemistry as a rather trivial application of quantum mechanics. He had not fully imagined what remarkable structures can be built by natural selection, nor just how many distinct types of proteins there might be.

It is clear Max felt that uncovering the secrets of genetic replication would involve major new ideas, something like quantum mechanics, the theory that dominated his intellectual youth. That theory had also appeared suddenly, but it took years after that to arrive at a satisfactory interpretation. And the interpretation arrived at in Copenhagen did not please everybody, as Max had witnessed in the continued debate between Bohr and Einstein. He expected that something similar, even if not quite so extreme, would occur in genetics. The very simplicity of Watson and Crick's model took Max by surprise. In a 1972 interview with Horace Judson, the author of *The Eighth Day of Creation*, he reminisced about how he felt in 1953 when he learned of the double helix model:

Nobody, absolutely nobody, until the day of the Watson-Crick structure, had thought that the specificity might be carried in this exceedingly simple way, by a sequence, by a code. This denouement that

came then—that the whole business was like a child's toy that you
can buy at the dime store"—Delbrück laughed and shook his head—
"all built in this wonderful way that you could explain in Life maga-
zine so that really a five-year-old can understand what's going on."
He laughed again, the same rueful way. "This was the greatest sur-
prise for everyone.

But, convinced of the double helix's importance for genetics, Max
swung into action. From his pulpit as Pope of the Phage Church, he
made sure people were aware of this *"greatest surprise"* and that a new
"tumultuous phase" of biology was beginning. The topic for the 1953
Cold Spring Harbor symposium was viruses. When Watson, coming
directly from England, arrived at the meeting in early June, he found
Max handing out copies of the original Watson-Crick paper and of a
second one the two had written a month later, spelling out the model's
implications for genetics. Max also scheduled a special presentation
for Watson to deliver his latest findings.

There were many questions in the discussion that followed Watson's
talk, but one stuck in Watson's mind. Leo Szilard, whom Watson described
to me as *"always thinking three or four steps ahead of everybody else in*
both science and politics," asked, *"Can you patent it?"* Watson knew this
was not an idle question, since Szilard had filed patents for refrigera-
tion mechanisms together with Einstein and had patented the idea of
a nuclear chain reaction in 1934, four years before fission was discovered. But patents are given only for useful inventions, and nobody at
the time could think of any use for DNA. Szilard then suggested the
alternative of taking out a copyright. More than a half century later,
the idea of patenting brief stretches of human DNA has become a
central issue in gene sequencing.

Specifying the structure of DNA was indeed a giant step forward,
but many problems remained, the most obvious of which was to determine how the molecule's stored information could be retrieved and

transmitted. This presumably required separating the intertwined chains so that some yet-to-be-specified chemical agent could read the genetic message encoded in the sequence of bases. Max believed it would not be possible for the two chains to be unwound without either breaking or getting tangled with each other. He made it clear that he regarded this problem as crucial and the model as unworkable until a way to solve this conundrum could be found. Watson and Crick felt otherwise. In their second *Nature* article they had already written:

> *Since the two chains are intertwined, it is essential for them to untwist if they are to separate. . . . Although it is difficult at the moment to see how these processes occur without everything getting tangled, we do not feel that this objection will be insuperable.*

A year later Max proposed a solution to this problem that involved cutting each of the two chains into pieces so that they could be separated and later stitching the pieces back together. He was wrong. No cutting and stitching was required, though it did take a few years before an extraordinarily beautiful experiment by two young Caltech researchers, Matt Meselson and Frank Stahl, proved conclusively that the solution was the one Watson and Crick had originally envisioned: the two chains simply unwind. As Watson writes in a recent book in which he discusses the events of the 1953 summer, *"Increasingly, by then, I and other younger members of the phage group were realizing that Max's scientific hunches were frequently dreadful. But once he saw the truth, he quickly and gracefully reversed course."*

It is also true that Max was preparing to leave phage genetics. He was feeling that molecular biology was in good hands and now becoming increasingly popular. This made it easier for him to depart, for he preferred following his own frequently delivered admonition *"Don't do fashionable research."* The Watson-Crick discovery had strengthened this resolve. Its very beauty and simplicity meant the field would grow quickly in popularity and certainly become *"fashionable."* The problem of genetic replication he had set out to solve had been dealt with, and now, in Geo-like fashion, he was ready to move to a new field.

What happened next, in a strange turn of events, is that as Max moved away from molecular biology, Geo moved toward it.

32 Geo and DNA

The point, the contribution of Gamow was that he made one realize
that there was—or that there might be—an abstract problem that
was independent of the machinery.

> *Francis Crick, on Geo's proposal for how the genetic code*
> *functioned*

During the early summer of 1953, while visiting the physics
research group in Berkeley, Geo ran across Luis Alvarez holding the
latest copy of *Nature* in his hand. Seeing Geo, Alvarez said to him,
"Look what a wonderful article Watson and Crick have written." Alvarez
was a tall, blond-haired, blue-eyed native Californian, a man widely
regarded as one of the twentieth century's most talented experimental
physicists. He was already famous for his work in cosmic rays, isotope
research, and radar, and for helping develop the implosion mechanism
used in the atom bomb dropped on Nagasaki. Though he would go
on to win the Nobel Prize in Physics in 1968 for building detection
devices used in elementary particle physics, he was better known to
the public in later years for his schemes to detect hidden chambers
in the Egyptian pyramids and for being the originator, together with
his geologist son, Walter, of the theory that the extinction of the dino-
saurs was due to the dust cloud injected into the earth's atmosphere
by a collision with a large meteorite. In short, Alvarez was the sort of
person who always had interesting ideas and was worth listening to.
Geo took the suggestion and read Watson and Crick's article.

On his way back to Washington from California, Geo wrote the authors a letter, introducing himself and telling them, *"I am very much excited by your article in May 30th Nature and think this brings biology over into the group of exact sciences."* He then asked a series of questions about how the sequence of bases in DNA might specify genetic information. Watson and Crick were not altogether sure that their new correspondent, already famous for his practical jokes, was serious. They soon received a draft of a short manuscript about DNA from him and realized that he was. Meeting Geo soon thereafter, they began to appreciate how stimulating it was to talk to him—and how much fun.

Geo had an idea about DNA. Like his idea about the origin of the chemical elements, it was wrong, but again like that earlier idea, it was *"interestingly wrong."* Indeed it was very interestingly wrong. While others were thinking of ways to test the double helix model, Geo simply assumed it was correct and proceeded from there. His thought was that if DNA was to specify genetic information, it had to be able to deliver the information on how to assemble amino acids, the building blocks of proteins. But how was this done? He provided an answer.

To see his solution, call the DNA bases 1, 2, 3, 4 instead of A, T, G, C, and draw them as equal-size circles. Using the circles as units, construct two chains, each of them wrapped around a cylinder, as in figure 3; these form a double helix. The two chains are also bound together because the bases on the one are linked to those on the other. The linking must, however, satisfy Chargaff's rule, which stipulates that A connect only to T, and G to C, or, equivalently, 1 connect to 2, and 3 to 4. Looking at figure 3, you see diamond-shaped openings on the side of the cylinder between the two chains. This was where Geo thought the amino acids would fit in, like keys in a lock, with each diamond determined by the four bases that surround it.

This scheme would work only if the amino acids could be squeezed into the diamonds. In order for this to be possible the amino acids had to be about the same size as the bases. A quick check of references showed Geo that they were. This cleared hurdle number one. In addition, though four bases labeled each diamond, one of them was predetermined because the tie across chains was such that 1 could link only to 2, and 3 only to 4. This meant that the "code" for an opening, or, in other words, for an amino acid, was a three-letter word made of As, Ts, Gs, and Cs, or, equivalently, a three-number sequence made of 1s, 2s, 3s, and 4s. The question, then, was how many three-letter words one could make with a four-letter alphabet, assuming permutations did not change the meaning. Look at figure 4, and you will see the answer is twenty: ten with 1 coupling to 2, and ten with 3 coupling to 4. Since permutations didn't matter, which of the two numbers was on top and which was on the bottom was irrelevant; to give an example, in the case of 1 linking to 2, the same amino acid was selected whether there was a 1 on top and a 3 on the bottom or vice versa.

Greatly excited, Geo now consulted biochemistry texts to see how many amino acids there were. The answer was apparently twenty-five, but some seemed to be questionable assignments. This was close enough. Geo began to think might have discovered the secret of how DNA made amino acids. And there was still more. Once the amino acids moved out of the diamond-shaped locks where they were formed, the order in which they were assembled oriented them like a

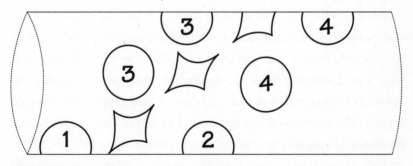

FIGURE 3
Model of how amino acids fit inside DNA chains

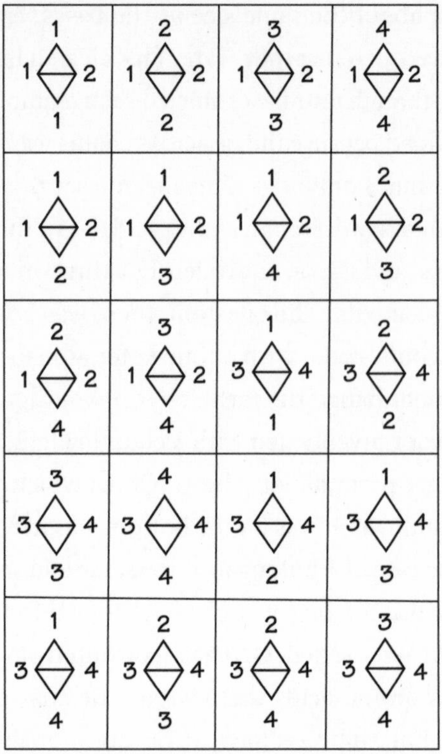

FIGURE 4
The twenty amino acids
of Gamow's scheme

string of beads. But this is exactly what a protein was. Geo's scheme explained how DNA coded for amino acids and how amino acids were joined into proteins. Might he, a rank outsider, have cracked a great problem in biology?

Being testable was another of his model's great virtues. You can see this by looking back at figure 3, which represents the double helix's two chains. Two of the diamonds, or amino acids, have 3 pairing with 4, and the third has 1 pairing with 2. Focus on the middle one in which 3 pairs with 4: I just said that the same amino acid would result whether 3 was on top and 2 on the bottom or vice versa. That remains true, but the sequence of amino acids would not be the same, because if 3 was on the bottom and 2 on top, the other circles would also have to be rearranged, so the coupling across chains remained 1 to 2, and 3 to 4. In other words, though all amino acids could be assembled, not all sequences of amino acids could. This imposed strict restrictions on protein formation. The proposal could be disproved! Geo wrote a very short note about all this, attached illustrations, and sent it off to *Nature*. They published it in February 1954.

He of course also sent a copy of his note to Watson and Crick. Reading it, they quickly came to the conclusion that Geo's list of amino acids was faulty. He had left out asparagine and glutamine and had included some dubious ones. But they also recognized they didn't know how many amino acids there were, nor had they realized that it was an important question to ask. The answer was not straightforward, for biochemistry books listed all possible candidates, even ones that appeared in only a few proteins. As Crick says:

> *The idea that there might be a standard set of amino acids and the rest were, in some sense, freaks had not penetrated to most biochemists, though obviously some protein chemists thought that way even if they had not formulated their ideas explicitly.*

By "*freaks,*" Crick meant amino acids that were synthesized from those on the standard list and therefore should not be counted or ones that occurred very rarely and only in special organisms.

Crick and Watson decided they had better make their own inventory of standard amino acids. When they finished, much to their surprise, they found the list had twenty members. They still didn't accept Geo's proposal, because they thought it highly unlikely that amino acids were assembled directly on the DNA, but the appearance of the number twenty was tantalizing. Recalling that period, Crick said:

> *The point, the contribution of Gamow was that he made one realize that there was—or that there might be—an abstract problem that was independent of the machinery. By thinking of it as an abstract problem of going from one thing to another, you might be able to deduce something about it. As we know—as it turned out, you*

*couldn't. You see. But nevertheless it drew attention to what it was
you had to discover—namely what you would now call the genetic
code. Which wasn't a thing most people realized might exist.*

Geo's note to *Nature* had been telegraphic. He now prepared a longer manuscript and submitted it to the National Academy of Sciences
for publication in its *Proceedings*. Having just been elected an academy
member and knowing that members' papers were not refereed, Geo
thought he could bypass the sort of criticism a professional society's
editors would normally make of such a bold proposal.

A few days later he received a phone call from Merle Tuve, a friend
who also happened to be the National Academy's home secretary. Tuve
asked him to withdraw the paper *"because, of course the rule is since
you are an elected member you can send anything you like and if you
send it in physics, it's another matter. But this is biology, and biologists are
unhappy about it."* One reason for the biologists' unhappiness might
have been seeing Mr. C. G. H. Tompkins (the imaginary character in
much of Geo's popular writings) listed as co-author. They suspected
Geo was making his debut in the academy's journal with a prank at
their expense, choosing them rather than his fellow physicists as victims. Though this was not Geo's intention, he acquiesced in the decision; the manuscript, stripped of its questionable co-author, was sent
to the Royal Danish Academy, of which he was also a member, and was
published in its *Proceedings*. Geo later sent reprints of it to biology
members of the National Academy, writing *"Best regards from Geo"* on
each one. The paper was not a prank.

Watson and Crick thought Geo's proposal for building amino acids
directly on the DNA molecule was probably wrong, because proteins
are assembled outside a cellular nucleus, while DNA resides for the
most part on the inside (which is why Miescher, DNA's discoverer,
called it nuclein). It seemed therefore likely that an intermediate step
would be needed between reading the genetic code on DNA and using

that information to construct proteins. If so, there had to be a carrier of information from one site to the other, and RNA was the logical candidate. But Crick was intrigued by the fact that Geo's model was testable, and he set to work trying to disprove it. It wasn't easy but within a few months, using the still-limited data available on amino acid sequences in proteins, he had shown that many of the variations of Geo's scheme could not hold.

Geo kept Bohr apprised of his thinking, as he had done consistently over the years, letting his old friend know that he was now working on biology problems. In a postscript to a 1954 letter to Bohr, he added, *"I am still nuclear physicist since nucleic not far from Nuclear, Proteins are close to Protons and there is only one letter different between H bonds and H bombs."*

But for the time being Geo was focusing on biology and not really thinking much about nuclear physics. Taking up residence in Berkeley in the winter of 1954 for a sabbatical term, he drove to Pasadena to visit Max and to meet Watson, who had in the meantime taken a position as a postdoctoral fellow in Max's group. Arriving in his brand-new white Mercury convertible with red leather upholstery and joking in his squeaky voice with a curious Russian accent, Geo hardly went unnoticed, as usual. He quickly made friends with the growing Caltech genetics group, especially with Watson and Alex Rich, a postdoctoral fellow in Pauling's group who was working with Watson on the problems of RNA, the molecule that was increasingly coming to be seen as a necessary component in explaining the connection between DNA and proteins.

Returning from his trip to southern California, Geo went to a Berkeley dinner party with Watson, Leslie Orgel, and Gunther Stent, all young molecular biologists interested in RNA's structure. Over the course of the evening this group had the idea of founding an association that would have twenty members, each of whom would be assigned an amino acid nickname; for example, Watson was proline and Geo was alanine. They would use the nicknames to sign letters to one another, sharing them if noteworthy; their subject was to be news about RNA. The group would be known as the RNA Tie Club.

The four participants at the dinner party were of course charter members of the club, as were Francis Crick, Alex Rich, and Max. Others included Richard Feynman and Edward Teller, both now interested in the question of a genetic code. Geo then took the joke of the RNA Tie Club a step further, by having a special silk tie made for members by a Los Angeles haberdasher. Accompanied by a tie pin adorned with the first three letters of the member's amino acid, the tie was meant to document membership in the club. Geo also designed stationery with a club motto suggested by Max: *"Do or die, or don't try."* The club's officers were recorded underneath it; George Gamow—Synthesizer, Jim Watson—Optimist, Francis Crick—Pessimist, Martinas Ycas (another young biologist)—Archivist, and Alex Rich—Lord Privy Seal.

Amusing stationery had always been a specialty of Geo's. It added to the pleasure of receiving one of his humor- and cartoon-filled letters. For personal matters he used a letterhead with a beautiful drawing of his house and garden. Beneath that, in Gothic script, it read, "The Gamow Dacha." His cosmology friends received letters that featured his address and, below it, in increasingly smaller print, "United States–North America–Planet Earth–Solar System–Milky Way–Virgo Cluster–The Universe."

With his interest in DNA and RNA continuing, Geo was excited to learn that Watson would be spending the summer of 1954 on Cape Cod at the Woods Hole Marine Biological Laboratory and that Crick would also be there for a good part of the time. Deciding he wanted to join them, Geo called a friend who lived in Woods Hole and, by return mail, received an invitation for the whole Gamow family—Geo, Rho, and eighteen-year-old Igor, then studying to be a ballet dancer—to spend three August weeks at his cottage by the water.

Crick described what life with Geo was like during those three weeks:

On most afternoons Jim and I went out to the cottage and sat on the shore with Gamow, discussing all different aspects of the coding problem, idly chatting or just watching Gamow show some of his card tricks to any pretty girl who happened to be around. The pace of scientific life in those days was less hectic than it is now.

It's not quite clear what *"less hectic"* meant. Crick and Watson were certainly not yet famous and did not have as many demands on their time as they would have later, but I think Geo always enjoyed showing his card tricks and playing practical jokes. Doing research and having quite ordinary fun seem to have been more intertwined for him than for almost any other major science figure.

Watson also described that period, including taking note of Geo's drinking: *"Particularly lethal were the drinks that Geo prepared, for his idea of a tall drink was a tall glass completely filled with whisky."* During these years all the young geneticists developed warm feelings about Geo. In *Genes, Girls, and Gamow,* written by Watson as a sequel to his *The Double Helix* more than thirty years later, a photograph of the young Watson sporting his tie from the RNA Tie Club appears on page one. The book concludes with more than twenty pages of what Watson calls "George Gamow Memorabilia," mainly letters to Tie Club members written in the mid-1950s. Aside from their intellectual content, these sui generis literary productions are embellished by Geo's skill as an illustrator. A discussion of the virus that attacks tobacco plants might be accompanied by a drawing of a happy smoker flicking cigarette ashes, and the question of what can be accommodated into the spacing between bases on a DNA molecule sidetracks Geo into contemplating what fits into the open mouth of a tiger (see figure 5 on page 218). In the end, however, the letters always came back to science questions.

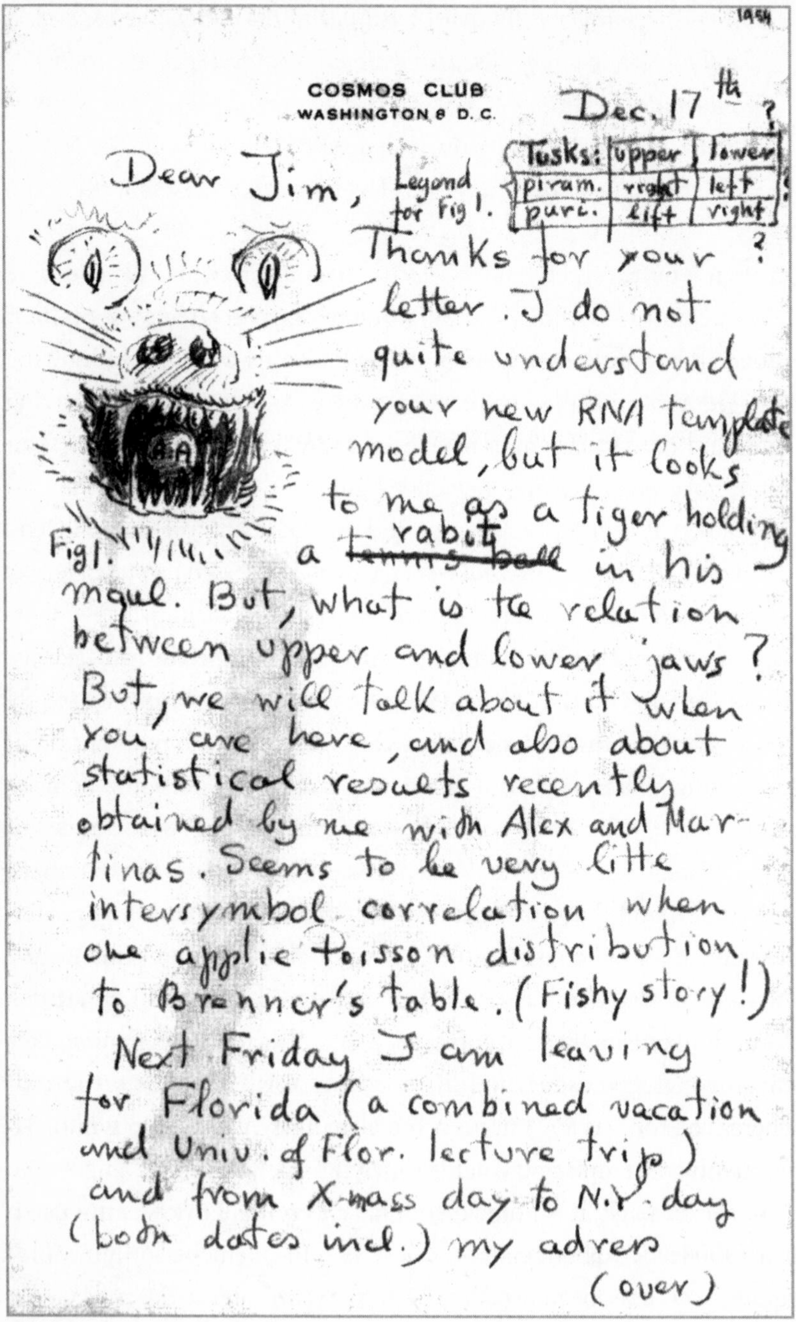

COSMOS CLUB
WASHINGTON 8 D.C.

Dec. 17th ?

Dear Jim, Legend for Fig 1.

Tusks:	upper	lower
piram.	rrght	left
pure.	lift	right

?

Thanks for your letter. I do not quite understand your new RNA template model, but it looks to me as a tiger holding a ~~tennis ball~~ rabit in his mgul. But, what is the relation between upper and lower jaws? But, we will talk about it when you are here, and also about statistical results recently obtained by me with Alex and Martinas. Seems to be very little intersymbol correlation when one applie Poisson distribution to Brenner's table. (Fishy story!)

Fig1.

Next Friday I am leaving for Florida (a combined vacation and Univ. of Flor. lecture trip) and from X-mass day to N.Y. day (both dates incl.) my adress

(over)

FIGURE 5
1954 letter to Watson from Geo

33 Geo Begins Again

The solution looks considerably less elegant than the simple theoretical correlation which I had originally visualized, but it has the
indisputable advantage of being correct, elegant or inelegant.

Geo, on his reaction to the genetic code

Though Geo's suggestion for amino acid construction was wrong, he focused the thinking of the many who laid the ground for our modern understanding of genetics. The reactions geneticists had to his model echoed in some ways those of cosmologists to his proposal about the origin of the chemical elements. This is no accident, since it reflects Geo's style of thinking: bold suggestions combined with the notion that the simplest solution is the correct one. He had envisioned all elements being built up in the early universe, mocking those who said otherwise as being like cooks who required three ovens, but it turned out that many ovens were needed. In genetics, Geo imagined amino acids being assembled directly on the DNA chain, whereas not only is RNA needed, but also three different kinds of RNA are involved in linking the acids to DNA.

In this context, it is noteworthy that the words Enrico Fermi used in a 1950 lecture about Gamow's notion of the elements' origin could equally well have been used by Crick five years later to describe Geo's work on DNA. Fermi said:

We must take note of the courage with which Gamow has attempted to construct a theory based on extremely determined hypotheses: the theory has failed, meaning that some of the hypotheses are wrong, but the result thus achieved is certainly more remarkable than what one could have obtained by a vague theory that could have explained many experimental facts because of its arbitrariness.

Geo's cosmology papers are correct enough to be hailed as classics in the field, while his genetics papers are not. However, if his model of amino acid formation had been right, it would have been regarded as one of the twentieth century's science landmarks. As it is, in his own words, *"The solution looks considerably less elegant than the simple theoretical correlation which I had originally visualized, but it has the indisputable advantage of being correct, elegant or inelegant."*

Though it is dangerous to generalize, physicists usually subscribe to a notion that is called Occam's razor, attributed perhaps apocryphally to the fourteenth-century Franciscan friar William of Occam. In essence, it says that the correct explanation for a complex set of phenomena is the simplest one.

Molecular biologists, on the other hand, feel it is wise to remember the two maxims attributed to Leslie Orgel, one of the RNA Tie Club's charter members. The first is *"Whenever a spontaneous process is too slow or too inefficient, a protein will evolve to speed it up or make it more efficient."* The second, even more to the point, says, *"Evolution is cleverer than you are."* By cleverness, Orgel meant that evolution stumbles upon solutions to problems by trying many avenues and is then quick to exploit them. They may not be the most elegant solutions, but, as Crick wrote, *"The genetic code, like life itself, is not one aspect of the eternal nature of things but is, at least in part, the product of accident."*

It took years to work out all the details, but the essential points of how DNA actually leads to proteins first appeared in a typewritten seventeen-page letter Crick sent to his fellow RNA Tie Club members in early 1955. It begins by laying out the reasons why Gamow-like proposals would almost certainly not work, but nonetheless concludes the first section by praising Geo: *"it is obvious to all of us that without our President the whole problem would have been neglected and few of us would have tried to do anything about it."*

Triplets of bases do specify amino acids, but, in contrast to what Geo thought, the order of the bases *does* matter—AGC codes for serine and ACG for threonine. There are therefore $4 \times 4 \times 4 = 64$ triplet possibilities, but many of them specify the same amino acid—that is, ACC and ACA both code for threonine. In addition, amino acids are assembled outside the cell nucleus, not on the DNA chain. Three separate forms of RNA (messenger, ribosomal, and transfer) are required to construct proteins. The progression begins when the two chains of DNA unzip totally or partially; a section is then copied (the process is called transcription) onto the first type of RNA that is needed, messenger RNA. This molecule leaves the nucleus, migrates into the cellular medium (cytoplasm), and attaches to a ribosome (two thirds ribosomal RNA), a spool-like unit where the messenger RNA information is read (this process is called translation) by small molecules of transfer RNA. Each of the latter has an RNA triplet on one end that recognizes its counterpart from the messenger RNA. On the other end it has an amino acid. Phrasing this somewhat differently, this specialized molecule *transfers* the genetic information to the protein construction one building block at a time. By repeating the process with however many amino acids are needed, a full protein is assembled, one that reflects the original DNA message. Occasionally, as in the case of retroviruses, Crick's central dogma of DNA to RNA to protein does not hold, and RNA leads to DNA instead of the other way around.

As these new developments became established, Geo's involvement in genetics waned. For a while he collaborated on the problems of amino acid construction with Alex Rich, who had moved in 1954 from Caltech to the National Institutes of Health in Bethesda. But in 1955 Rich took a leave to work in Cambridge, England, with Francis

Crick and Jim Watson. This formidable trio's aim was simple: decipher RNA's structure. Other groups were trying to do the same, but the Cambridge three were quite sure they knew more about the subject than anybody else.

A few days after Rich had arrived in England, Watson received a letter from Geo letting him know that the three of them had been scooped. The letter had a brief introductory paragraph, and then Geo asked Watson:

> What do you think about this Rundle's paper? Should we elect him as an honorable member of the club or should we disband the club altogether? I think I will go back to cosmology! I think the most interesting thing about this RNA model is that it explains that all proteins have 3N amino acids.

An hour later, an upset Rich came into the laboratory. He, too, had received an airmail letter from Geo telling of Rundle's work; he also had a report of the discovery from Max in Pasadena. This read:

> Here of course everybody is buzzing about Rundle's RNA structure (JACS last issue). It seems the most unbelievable thing that a complete outsider from the Middle West should have hit the jackpot, and that he should have kept it to himself during all the time the paper has been in press. . . . Anyhow, since the biochemical meaning of the structure is almost more obvious than the Watson-Crick structure, it would be hard to convince anybody that the structure is wrong.

JACS was the Journal of the American Chemical Society. Crick, Rich, and Watson quickly found out, by consulting Cambridge colleagues in chemistry, that the latest issue had not yet reached England (ships were still the standard means of conveying journals across the Atlantic then). Robert Rundle, an inorganic chemist at Iowa State

University, was an unlikely candidate to have solved the RNA puzzle but, as a former co-worker of Linus Pauling's, not an inconceivable one.

After several hours of shock, it dawned on the trio that Rundle's scooping them might be a hoax, a coordinated joke played on them by Geo and Max. Summoning up his courage, Rich placed a transatlantic phone call (still a rarity in those days) to a chemist friend in Washington, D.C., who revealed to them that Geo and Max had concocted the scheme on a recent visit of Max's to Washington. Twenty-five years after their time together in Copenhagen, Max and Geo could still pull off a good joke. The Cambridge group may have been amused at having been played, but I suspect their dominant emotion was relief.

34 Max Begins Again

It was you who inspired me thirty years ago to go into biology and
I believe I am the only one of your disciples who has made his
way in this direction.

Max, in a 1962 letter to Niels Bohr

By 1956 Geo was ready to leave biology and start over again
in science. His marriage to Rho had also come to an end, and he felt
a change of scenery would do him good. When he received an offer
of a professorship from the University of Colorado at Boulder in the
summer of 1956, he accepted without hesitation. Max also moved in
1956, though while Geo was going west, he was going east, all the
way to Germany.

In the mid-1950s Max had been experiencing some new beginnings
as well, though luckily for him they did not involve divorce. Despite
their having very different backgrounds, Max and Manny enjoyed a
marriage that was singularly happy and harmonious. It was much
admired by friends and colleagues, all of whom greatly appreciated
being drawn into the greater Delbrück family.

Max's new beginnings were purely professional. In 1954 he had
succeeded in doing something that was important to him: becoming involved once again with academic life in his old homeland. This
would be his first semiofficial visit to Germany and he chose Göttingen, his alma mater, as the place to deliver a set of lectures on the

new genetics. Though German biology departments were still organized around the classical subjects of zoology and botany, German universities were aware that molecular biology was emerging as a discipline. They knew about the discovery of DNA's structure and were interested in what it meant for their future. Göttingen hoped Max would show them how to join the upcoming science revolution.

They had not, however, anticipated that Max would administer shock therapy by exposing his audience to the kind of informality in science discussions that had now become the norm at places like Caltech and Cold Spring Harbor. When Göttingen's dean of natural sciences arrived at Max's inaugural lecture to extend his university's welcome, he was dressed in his official robes despite the early summer heat. By contrast, Max was clad in a T-shirt and sneakers; he quickly thanked the dean and then announced that he would first show all the lecture's slides without explanation so that windows could be quickly opened and the room cooled off. This was not the kind of behavior Göttingen expected from the son of the great historian Hans Delbrück and the nephew of the equally great theologian Adolf von Harnack.

Some were appalled by Max's manners and his curtness, but the University of Cologne's Joseph Straub thought Max was the ideal person to lead the way into modern genetics research. Straub tried right away to lure Max back to Germany on a long-term basis. Max was not interested but agreed to come to Cologne in 1956 to deliver a three-month lecture series on phage genetics. When he and Manny and their two small children, Nicola and Jonathan, then aged eight and nine, arrived there, they found a city still in the process of rebuilding after its almost total destruction in World War II.

My own mother is from Cologne, and I remember as a child in the early 1950s going there with her from Italy to visit my uncles and aunts still living in the city. Cologne had been the first victim in World War II of what came to be known as saturation bombing, the dropping of ordinary and incendiary bombs on a fixed target by an armada of hundreds of planes flying overhead. Though I had seen a fair amount of postwar destruction in Italy, nothing prepared me for Cologne, the ruins made even starker by the city's gigantic cathedral seemingly untouched amid the rubble.

However, by 1956, when the Delbrücks arrived in Cologne, the German economy was reviving and the city's rebuilding was proceeding apace. The family enjoyed their stay, and Max now began to entertain the idea of returning to Cologne for a longer period. He announced that he would come for a two-year residency but only if certain seemingly impossible conditions were met. He insisted that an American-style institute had to be created, one where faculty members were equals and where students and postdoctoral fellows were free to move from one lab to another and to challenge their superiors. He wanted it to be a radical departure from the German university system.

It took years to carry the plan through to completion and to obtain the necessary funding, but the Kölner Institut für Genetik (Cologne Institute for Genetics) was ready to begin functioning in 1961 and the Delbrücks came back. By now there were three children; Tobi, thirteen years younger than Jonathan, had been born the year before the family arrived in Cologne. Ludina, thirteen years younger than Nicola, was born later, while they were there. Nicola remembers that she and her older brother, now teenagers, were sent to a local school and that

We all spoke German from the day we arrived in Germany and in a year I was thinking in German rather than in English. Instead of camping trips we took long Sunday walks in the woods surrounding the city. On very cold days we went ice-skating along the canal or climbed the cathedral tower.

This sounds like a normal German family but of course they were not, and Max never did become a "normal" German professor. Rainer Hertel, later a biology professor in Freiburg, remembers opening the door for him at their first meeting, bowing his head, and saying,

Guten Tag ("Good day"), Herr Professor Delbrück. Max's reply was *Diese Unterwürfigkeit ist grässlich ("This obsequiousness is horrible").*

By 1962 the Cologne Institute was ready for a grand official opening. The choice of inaugural speaker was evident. Writing Bohr from a vacation resort he had last visited thirty years earlier with Pauli and Gamow, Max told Bohr, now seventy-six, that his speaking in Cologne *"would seem wonderfully fitting. It was you who inspired me thirty years ago to go into biology and I believe I am the only one of your disciples who has made his way in this direction."*

Bohr gladly accepted Max's invitation. On June 21, 1962, he delivered an address in Cologne. It was entitled, at Max's suggestion, "Light and Life, Revisited." In it Bohr said that thirty years earlier he had suggested *"that the very existence of life might be taken as a basic fact in biology,"* but the developments in the intervening years now led him to believe that there was no need for any *"limitation in the application to biology of the well-established principles of atomic physics."* The question that had set Max off on his quest thirty years earlier was being put to rest.

This was Bohr's last public speaking appearance. A few days later he had a minor cerebral hemorrhage and, on November 18 of that same year, a fatal heart attack. It was the end of an era in science and of personal and intellectual relationships that had done so much to shape Geo's and Max's lives. Bohr had guided their careers at an early stage, but even more than that, they had shared both his passion for science and his joie de vivre. They had reinterpreted both in their own fashion, adapting them to their new environments, but the Copenhagen spirit had stayed with them.

35 The Molecular Biology That Was

Max was a surgeon of the mind. He asked for an experimental result
and then posed a series of questions that cut through to the heart
of the matter. For Max, science was a kind of intellectual game that
one plays against oneself with one's own hands.

Enrique Cerdá-Olmedo, describing Max

It is traditional to mark a major scientist's sixtieth or sixty-
fifth birthday by a one- or two-day symposium that combines serious
scientific talks with retrospective remarks about the person being hon-
ored. These events are called Festschrifts, loosely translated from the
German word for "celebration publications." Max's was held at Cold
Spring Harbor in May 1966. Noting that this was also the twenty-first
birthday of Max's first Cold Spring Harbor phage course, the editors
marked that coming of age as well.

The proceedings, published in a volume entitled *Phage and the
Origins of Molecular Biology* (Max referred to it as Patoomb), consist
of more than thirty essays. They roughly trace the historical development of molecular biology and of Max's influence up until 1966.
Though the double helix model of DNA had been proposed only a
little more than a dozen years earlier, many of the Festschrift participants seemed to be thinking that the golden age of molecular biology,
or at least that their own participation in its unfolding, was coming to
an end. This was not for lack of interesting problems in the subject,

but rather, accustomed to being a small cadre, they were uncomfortable seeing so many scientists engaging in the type of research they had pioneered. As Luria commented somewhat wistfully, it was now true that *"annual phage meetings at Cold Spring Harbor attract literally hundreds of participants."*

Seymour Benzer, an inventive physicist who had turned biologist, provides one example of the pioneers' mood shift, describing in Patoomb his own transition away from molecular biology. He felt that his work had become *"more and more exciting, until it dawned on me how many other people were doing the same things. I had almost gone down the biochemical drain. Delbrück saved me, when he wrote to my wife."*

It was only a footnote in a letter from Manny to Benzer's wife, Dotty, but the message from Max was clear:

> *Dear Dotty: please tell Seymour to stop writing so many papers. If I gave them the attention his papers used to deserve, they would take all my time. If he must continue, tell him to do what Ernst Mayr asked his mother to do in her long daily letters, namely underline what is important.*

Benzer heeded the message and set off in a totally different direction, now looking for an organism in which he could identify genes affecting behavior. He soon settled on fruit flies, *Drosophila,* and so began a long Caltech career in which he re-created a modern version of Morgan's old Fly Room.

Having more or less left research in molecular biology by the mid-1950s, Max was the first of the pioneers to have done so, departing at the beginning of what many thought was the subject's golden age, the years following the discovery of the double helix. Twenty years earlier he had started abandoning physics in what many thought of as *its* golden age, in order to study the identity of a gene, how it replicates and how it mutates.

Max had seen quantum mechanics being led by an emerging group of youngsters; in 1953 he began to witness the same phenomenon in molecular biology. These new leaders were twenty or so years younger than he, just like the "boys" of quantum mechanics had been twenty years younger than Bohr. But even though Max had in many ways adopted at Caltech and Cold Spring Harbor the lessons he learned in Copenhagen, he did not play a role in molecular biology's progression like Bohr's in quantum mechanics. Rather than attempting to guide the young in their explorations of genetics, Max felt an inclination to move to another field.

I think part of the urge he experienced came from his sense that molecular biology would develop in ways in which biochemistry, never his forte, would play an increasingly important role. As Alex Rich once said to me with a smile, *"Max lost interest in molecular biology when it became molecular."*

I also think that at heart Max remained a physicist or, at the very least, enjoyed the ambiguity of whether he was a physicist or a biologist. He had written Bohr in January 1947 that his acceptance of the Caltech offer marked the completion of his metamorphosis into a biologist. But three years later, in 1949, when he joined the composer Paul Hindemith and the poet Wallace Stevens in giving talks at a jubilee meeting of the Connecticut Academy of Arts and Sciences, he chose to entitle his address "A Physicist Looks at Biology." Even two decades after this talk, accepting the 1969 Nobel Prize in Physiology or Medicine, he spoke on "A Physicist's Renewed Look at Biology: Twenty Years Later." This leads me to think that, even in his own eyes, Max's metamorphosis was never complete.

Clues to why Max elected to continue calling himself a physicist have little to do with the problems he was working on; his research lay solidly in the realm of biology. And yet he continued to say that he thought like a physicist. But what did that mean? Clues come from the distinction between physics and biology he made in his Connecticut talk.

Max viewed living organisms as being classified on a continuum, each at least slightly different from the others, but all their existences woven together in the fabric of evolution. Accordingly, he thought that biology has no *"absolute phenomena,"* a term he used to identify occurrences that take place over and over again in exactly the same manner. By contrast, he believed that physics features an abundance of such events. Every hydrogen atom, whether in a glass of water or on a stellar surface, is identically equal to every other one, and the laws that govern their behavior never change. *"Absolute phenomena"* made it easier to identify what Max called *"clear paradoxes,"* things that did not fit into the accepted schemes. As is evident in his 1949 address to the Connecticut Academy, the search for such paradoxes is what had drawn him to biology:

> *In biology we are not yet at the point where we are presented with clear paradoxes and this will not happen until the analysis of the behavior of living cells has been carried into far greater detail. The analysis should be done on the living cell's own terms and the theories should be formulated without fear of contradicting molecular physics. I believe that it is in this direction that physicists will show the greatest zeal and will create a new intellectual approach to biology which would lend meaning to the ill-used term of biophysics.*

But, much to Max's surprise, the mechanism for genetic replication that followed from the structure of DNA did not present any *"clear paradoxes,"* nor did it require new laws of either physics or chemistry. The denouement seemed, as he put it in his 1972 interview with Horace Judson, to be *"like a child's toy that you can buy at the dime store."* Max now felt he would have to look elsewhere to find something in biology that would present him with a *"clear paradox."*

He chose sensory physiology for the quest, beginning by asking how simple organisms responded to stimuli. This was an interesting choice, worthy of the kind of deep thinking and precise organization that characterized his approach to scientific problems. He proceeded in much the same way he had done with phage, gathering a small group around him that included students and visitors from around the world, all engaged in the same kind of free-ranging discussions that had always marked his work. As one of those visitors, Enrique Cerdá-Olmedo, described him:

> *Max was a surgeon of the mind. He asked for an experimental result and then posed a series of questions that cut through to the heart of the matter. For Max, science was a kind of intellectual game that one plays against oneself with one's own hands.*

Sensory physiology occupied much of Max's attention during the remainder of his research career, but it proved less amenable to clear-cut answers than bacterial genetics. The organism he chose to study, *Phycomyces,* a fungus that sprouts large straight-standing stalks that are sensitive to light, touch, and other disturbances, gave no clear answers. The results were very interesting but did not, as Max had hoped, lead to finding the hydrogen atom of behavior.

But Max's quiver had many arrows, and the fungi were not his only targets. Since all responses to stimuli ultimately depend on messages being transmitted along nerve fibers, Max began to think about how this might take place. Geo had brought together physicists and astronomers, with sallies into biology. Max, who had joined physicists and biologists, decided it was time to bring computer scientists into neurobiology. This breadth of thinking was one way in which he and Geo were far from "ordinary."

Carver Mead, a professor emeritus of engineering at Caltech, is one of the world's pioneers in microelectronics, spearheading silicon chip development, VLSI (very large scale integrated) circuits, and many other innovations. But in 1967, when he was only a young Caltech faculty member, Max went to see him to inquire what sort of similarity there might be between nerve membrane channels and transistors.

This was the beginning of a new approach to neural transmission. After his meeting with Mead, Max did continue to do some work on membranes for a number of years, but his main function was to steer others into what he saw as a promising field, by providing guidance and support.

Eventually Mead would become the leader in the field known as neuromorphic engineering, which aims to construct models of sensory functions in the body, as for instance the building of a silicon retina. This is now a special interest of Max's son Tobi, a researcher at the same Zurich university Max went to eighty years ago to work on quantum theory with Wolfgang Pauli.

36 The Phage Church Trinity
Goes to Stockholm

One is what one is, partly at least.

From Molloy, by Samuel Beckett

I was more thrilled by the award of the Nobel Prize to him than about the award to me.

Max, on his reaction to Beckett's Nobel Prize in Literature

Max's research after the mid-1950s did not have the impact of his earlier work; his later years were, however, marked by much recognition, including the 1969 Nobel Prize in Physiology or Medicine, which he shared with Hershey and Luria, the other two members of the Phage Church Trinity. Max was pleased but not overwhelmed; on the day of the notification he still played the tennis game that had been scheduled for the afternoon. He was excited, however, to discover that Samuel Beckett, a writer he had read and greatly admired, would be getting the Nobel Prize in Literature at the same time. The intensely private Beckett, only six months older than Max, did not make the trip to Stockholm. At the very end of his own acceptance speech, Max commented on Beckett's absence:

While the scientists seem elated to the point of garrulousness at the chance of talking about themselves and their work, Samuel Beckett, for good and valid reasons, finds it necessary to maintain a total silence with respect to himself, his work, and his critics. Even though I was more thrilled by the award of the Nobel Prize to him than about the award to me and momentarily looked forward with intense anticipation to hearing his lecture, I now realize that he is acting in accordance with the rules laid down by the old witch at the end of a marionette play entitled "The Revenge of Truth."

The *"marionette play"* Max refers to takes place as an episode within the short story "The Roads Round Pisa," by the Danish writer Isak Dinesen, another of Max's favorite authors. This is a tale of an early-nineteenth-century melancholic young Danish count who becomes embroiled in a series of curious encounters near Pisa. The play referred to in the story is one that begins with a witch pronouncing a curse on a house so that *"any lie told within it will become true."* At the play's end the witch reappears and gives a little speech, which Max quoted verbatim as the conclusion of his Nobel Prize acceptance:

"The truth, my children, is that we are all of us acting in a marionette comedy. What is more important than anything else in a marionette comedy is keeping the ideas of the author clear. This is the real happiness in life and now that I have at last come into a marionette play, I will never go out of it again. But you, my fellow actors, keep the ideas of the author clear. Aye, drive them to the utmost consequences."

It is a very curious finale for a science Nobel Prize address, quite unlike any other I have read. What message is Max trying to impart, and how does this explain why Beckett chose to avoid the Nobel Prize ceremonies? Reading it, however, helped me to understand why a number of individuals who knew Max well, while praising him for his great logical and critical ability, also spoke of him as the Zen master of molecular biology. This title has the traditional meaning of a wise man, one who instructs his disciples, which he certainly was, but it also carries the association of speaking in koans, paradoxical riddles with no clear meaning. A typical example is:

> *A master who lived as a hermit on a mountain was asked by a monk, "What is the Way?"*
> *"What a fine mountain this is," the master said in reply.*
> *"I am not asking you about the mountain, but about the Way."*
> *"So long as you cannot go beyond the mountain, you cannot reach the Way," replied the master.*

Consistent with the sayings of a Zen master, the necessity of *"keeping the ideas of the author clear"* might have been Max's explanation for why Beckett did not come to Stockholm, but perhaps what he was saying was no more than a koan.

Three years later, in 1972, Max returned to the writings of Beckett, citing his work in an essay entitled "Homo Scientificus According to Beckett." In it Max makes the comparison between scientists and Molloy, the mysterious protagonist of Beckett's novel by the same name, the man who mysteriously utters, *"One is what one is, partly at least."* Over the course of a hundred-page monologue, the vagrant Molloy describes lonely wanderings across a strange landscape. At one point in the description of his ramblings, he says, *"in order to blacken a few more pages, may I say I spent some time at the seaside, without incident."* The lonely beachcomber sits by the shore, admiring stones he collects, sucking on them, and transferring them from pocket to pocket of his jacket and trousers. Max felt that much of what scientists do is akin to this sifting of stones, though he is willing to grant that, unlike Molloy, they are also *"playing animals. They not only play alone, but they also play together, and if they are not too morose, they actually prefer to play together."*

He may have been thinking of himself or of Bohr as he wrote these words, but I wouldn't be surprised if a picture of Geo flashed across his mind. It is hard to find a better illustration of a scientist as a *"playing animal"* than Geo, a man for whom everything was a kind of game based on collecting the prettiest stones, ones that others did not seem to be able to see. Admittedly Geo might have needed a glass of scotch or vodka by his side as he sat on the beach examining what he had collected, but his laugh upon finding a new beauty would have been contagious, enticing others to follow his lead.

37   The Triumph of the Big Bang

The presence of thermal radiation remaining from the fireball is to be expected if we can trace the expansion of the universe back to a time when the temperature was of the order of ten billion degrees.

From an article by Robert Dicke et al.

Though Geo would not witness the great modern era of experimental cosmology, he did get to see others admire the prettiest and most valuable of the stones he had collected. The story of finding those stones is a case study of science's frustrations, triumphs, and serendipitous events. It involves a discovery made but not verified until years later, parallel efforts, competition, communication, and miscommunication. But in the end there was an advance that all could celebrate. What, in mocking Geo's conjecture, Fred Hoyle had derisively called the Big Bang has become the accepted picture of the universe's origin, and understanding it has justly been counted as one of the twentieth century's greatest scientific achievements.

The discovery of the radiation that indicated the presence of a Big Bang bears some resemblance to the uncovering of DNA's structure in that it was made by two young unknowns and came as a great surprise to the outside world. There was also an element of competition with a more established group, although in the case of cosmology, unlike Watson and Crick in their race against Pauling, the radiation's discoverers, Arno Penzias and Robert Wilson, were not aware they were competing with anybody. But the biggest difference was that, whereas Watson and Crick knew exactly what they were looking for, Penzias and Wilson did not set out to find what would eventually make them famous.

In 1963 they had begun working with a radio telescope that had a twenty-foot horn antenna. It was located at Bell Laboratories' Crawford Hill communications center, a few miles from the company's principal research center in Holmdel, New Jersey. The instrument Penzias and Wilson were using had been built three years earlier for connecting to the recently launched *Echo I* satellite. But in 1962 a better telecommunications satellite, *Telstar,* was placed in orbit, and the telescope was not needed anymore for commercial purposes. It was now available for any experiment that seemed reasonable to both staff and management.

At the time, Bell Laboratories had a strategy of sponsoring pure as well as applied research. The idea behind this plan was that bright scientists, freed to do any kind of research they wanted, might develop marketable instruments while pursuing their own goals of advancing the frontiers of science. Furthermore, even if nothing profitable resulted, the favorable publicity Bell gained by supporting exploration seemed to be worth the investment. The laboratory's discovery of the transistor fifteen years earlier is often cited as a prime example of the wisdom of this policy. Unfortunately, this philosophy, which richly rewarded scientifically and commercially for several decades, has all but disappeared following the breakup of the Bell monopoly and the increased pressure on industrial laboratories for quickly demonstrable gains.

Back in 1963, however, the strategy was in full bloom, so Penzias and Wilson quickly gained approval for their proposal to redeploy the no longer commercially needed radio telescope. They planned to use it to study signals emitted by stars within our galaxy, the Milky Way. The optimal wavelength for their intended reception was in the microwave range, a span that extends from roughly a tenth of a centimeter up to 30 centimeters, the latter being the width of a microwave oven. Penzias and Wilson's first order of business was eliminating background noise, the kind of buzzing one often experiences in radio signal reception. They chose to measure at a wavelength of 7.35 centimeters (a little less than 3 inches), because the sky was reportedly very quiet at this value. But, as they soon discovered, it wasn't as quiet as they had expected. Pointing the telescope away from the plane of our galaxy to eliminate other activity, the two young radio astronomers still detected an appreciable signal. It was present even if they moved the telescope around or changed the time of day the measurement was taken. Penzias and Wilson, thinking this was due to a defect in either the circuitry or the antenna, set to work on eliminating the background noise. However hard they tried, though, that buzz would not go away.

They took the antenna apart and rebuilt it, wrapped aluminum tape around electrical connections, removed two pigeons that had occasionally used the antenna as a resting place, and then scrubbed the antenna free of the deposited pigeon excrement. They compared the signal from the antenna with one from a so-called cold load, a specially prepared emitter cooled to near-absolute zero. By switching back and forth from it to the antenna, they hoped to discover background noise originating in their amplifiers. But no matter what they did, the buzz persisted.

By 1965 Penzias and Wilson had come to feel that they were detecting a real signal. Something seemed to be generating noise in their antenna in a way that was independent of the direction they pointed the telescope, the time of day, or the season of the year. They had no idea what the signal meant but concluded that it was not due to their instruments' malfunctioning.

Alpher, Gamow, and Herman would have known what the buzz meant, but nobody asked them. A group of four researchers at Princeton University, only thirty miles from Penzias and Wilson's research station, also knew what the buzz meant, but they, too, were unaware of what their neighbors were doing.

Forty-nine-year-old Robert Dicke, a well-known physicist with an unusual mix of talents and interests, was the Princeton group's leader. His Ph.D. thesis had been in nuclear physics, but during World War II he had worked on the development of radar and had even designed an antenna suitable for reception in the microwave region. After the war, now a Princeton faculty member, Dicke had demonstrated a flair for both instrument design and attacking questions of basic theory, with recent interests centering on problems in cosmology. This had led him to expect, much as Alpher, Herman, and Gamow had, that radiation from early events in the universe's history should be present if indeed the universe began with a Big Bang. Deciding to look for this radiation with the type of antenna he had developed during World War II, he convinced two young Princeton experimentalists, Peter Roll and David Wilkinson, to join him in building the necessary apparatus. They constructed a telescope designed to take a measurement at a wavelength of 3.2 centimeters.

Dicke also urged a young Princeton theorist, Jim Peebles, to make a thorough study of the kind of radiation that would be expected if the universe had undergone a Big Bang. By early 1965 Peebles had a manuscript ready. He had done the same sort of calculation that Alpher and Herman had carried out fifteen years earlier, but more thoroughly and with better data. Most important, he did it with a full and correct grasp of all the physical processes that contributed to the production of the radiation.

The Princeton group, assured by Peebles's calculations of the importance of what they were seeking, realized they were embarking on a great venture. This was nothing less than finding possible evidence that the Big Bang had taken place.

A chain of coincidences connecting them to the Bell group was now set in motion. In February 1965 Peebles gave a talk about his results at Johns Hopkins University. Kenneth Turner, a friend from graduate school who was in the audience, afterward mentioned Peebles's results to Bernard Burke, then his colleague at Washington's Carnegie Institution. Burke knew Penzias and, in the course of a telephone conversation with him, suggested that he contact Dicke about the noise coming from his antenna. Penzias did so, and the Princeton group was quickly on its way to Crawford Hill. It did not take them long to realize what Penzias and Wilson had discovered and to appreciate that they had been scooped. Radiation from the early universe had already been detected!

Dicke and his collaborators proposed to Penzias and Wilson that they submit back-to-back articles for publication, with the Bell group describing the observation and the Princeton one reporting on a theoretical interpretation of the data. In the first, Penzias and Wilson cautiously stated, *"A possible explanation for the observed excess noise temperature is the one given by Dicke, Peebles, Roll and Wilkinson in a companion letter in this issue."* The Princeton group was more assertive: *"The presence of thermal radiation remaining from the fireball is to be expected if we can trace the expansion of the universe back to a time when the temperature was of the order of ten billion degrees."*

Geo, Alpher, and Herman were elated but also miffed when they saw the twin articles—elated by the finding and miffed that they had not been given more credit for anticipating the discovery. The Princeton group did include a reference to them—*"this was the type of process envisioned by Gamow, Alpher, Herman and others"*—but Geo and his collaborators felt the acknowledgment should have been more detailed and generous. After Geo received a prepublication draft of Penzias and Wilson's paper, he wrote them, *"It is very nicely written except that 'early history' is not quite complete."* After going on for some time about the estimates Alpher, Herman, and he had made, Geo concluded, *"Thus you see the world did not start with almighty Dicke."* I doubt if Geo's resentment lingered, but Alpher and Herman continued to feel embittered by their perceived lack of recognition. My sense is that both the scientific community and the popular press have satisfactorily acknowledged their contribution. If anything, I feel sympathy for Dicke and his group, who missed by a hairbreadth making what may be the cosmological discovery of the century.

38   The Cosmic Microwave
Background Radiation

**I think most importantly, the "big bang" theory did not lead to a
search for the 3 degree Kelvin microwave background because it was
extraordinarily difficult for physicists to take seriously any theory
of the early universe.**

> *Steven Weinberg, on why a search for the primordial radiation
> was not undertaken until 1965*

Geo, Alpher, and Herman had proposed in 1948 a picture of
the universe as they imagined it to be moments after its creation. They
saw it as a dense expanding mixture of neutrons, protons, electrons,
and photons, all colliding with one another in such a way that the
whole was at one temperature. As the universe grew in size, it also
cooled, reaching in a few minutes a temperature low enough—still
billions of degrees—that helium nuclei formed. Geo, Alpher, and Herman had thought that heavier nuclei would appear next, but this did
not happen, and therein lay their failure. The overall picture of cooling and expansion, however, was correct.

We now believe that, propelled by the initial explosion, the universe continued to get bigger. It also went from billions of degrees
to millions and finally to thousands, but still in thermal equilibrium,
albeit at falling temperatures. Until then it had been too hot for electrons to attach themselves to protons or helium nuclei, but by a little

less than four hundred thousand years after the Big Bang, the universe had cooled to some three thousand degrees. It then reached a turning point, at which hydrogen and helium atoms formed.

Atoms, normally containing equal numbers of electrons and protons, are basically electrically neutral, because protons and electrons have an equal and opposite electric charge. Photons, electrically neutral themselves, do not interact appreciably with other neutral particles, so those primordial photons essentially ceased having contact with their surroundings once atoms formed. From then on they moved unhindered through space. Atoms continued to interact with one another through gravitational forces, altering their respective motion, as they coalesced into stars, galaxies, and other configurations. Matter clumped, but photons, still reflecting their early distribution, did not. An observer would therefore expect to see photons appearing from all directions, carrying with them a record of the condition they had been in when atoms appeared. These photons would provide a uniform background to every astronomical sighting.

This is exactly what Penzias and Wilson detected. Since their telescope functioned in the microwave region, their observation came to be known as the cosmic microwave background radiation, or simply CMBR. They had unknowingly been studying a picture of the universe a little less than four hundred thousand years after the Big Bang, the last instant when thermal equilibrium prevailed. It was truly a momentous discovery.

Since the finding is so important, it seems natural to ask why it took so long to begin the search. Even Penzias and Wilson's discovery was a fluke. They deserve a great deal of credit for trying again and again to determine the source of the signal they were finding rather than simply dismissing it, but they were certainly not looking for the cosmic microwave background radiation. Dicke and his collaborators were, but the experiment they planned could probably have been carried out a decade earlier. Why wasn't it? Why didn't Geo suggest that a search be mounted for that primordial radiation?

He considered in 1946 the idea of a rapidly expanding and cooling universe, but the prediction of a residual radiation temperature did not appear until the 1948 paper by Alpher and Herman, the two disciples/friends he referred to, depending on his mood, as his co-conspirators, co-cosmogonists, or co-creators. Though their estimate was five degrees above absolute zero, the closeness to the presently measured temperature of three degrees (see endnote for an explanation of why radiation emitted at three thousand degrees is presently seen at three degrees) is fortuitous, because at the time, the universe's average mass density, a necessary input in the calculation, was thought to be quite different from what we now believe. As data and techniques improved, however, Alpher and Herman persisted in attempting to obtain more credible values, publishing at various times during the following years values that ranged from a few degrees to a few tens of degrees. But nobody sought to build an experiment to look for the radiation. Why not?

There is no simple answer. The physics community's perception of Geo was probably one factor. He had started by trying to build a theory that would explain the abundance of all the elements; once it proved unsuccessful, there was a tendency to discount the whole argument rather than say that it worked for hydrogen and helium but not for the heavier elements. The different values predicted for the radiation, and perhaps the general feeling that Gamow was brilliant but often a little sloppy, may have contributed as well. His predictions were viewed skeptically, and his co-cosmogonists suffered by association. It is also true that communication between experimenters and theorists was poor, unlike what is now the case in cosmology. The former didn't know to look for the radiation, and the latter did not suggest they try to do so.

In *The First Three Minutes*, however, Steve Weinberg claims, I think correctly, that these aren't the main reasons the search wasn't mounted earlier. After discussing at length possible reasons for the delay, including the ones I list here, he writes:

I think most importantly, the "big bang" theory did not lead to a
search for the 3 degree Kelvin microwave background because it was
extraordinarily difficult for physicists to take seriously any theory
of the early universe (I speak here from recollections of my own atti-
tude before 1965). . . . The first three minutes are so remote from us
in time, the conditions of temperature and density are so unfamil-
iar, that we feel uncomfortable in applying our ordinary theories of
statistical mechanics and nuclear physics.

In other words, scientists in the 1950s felt *"uncomfortable"* thinking about anything as remote as the universe's first minutes. Geo was
different. Part of his special talent in science was feeling comfortable in places where others did not, in contemplating extreme conditions and in attempting to lead the rest of the physics and astronomy
community to do so. But we—that is to say, the rest of us in the
community—needed more than Geo's reassurance; we needed a sign
that this was the right path, and this is what the Penzias-Wilson measurement gave us.

Matters always seem much clearer in hindsight. Avery's studies of
pneumonia bacteria indicated unambiguously that DNA is the carrier
of genetic information, but it still took almost a decade and numerous other experiments before this information was fully appreciated.
There was a decade between Bohr's atom and quantum mechanics,
a decade between Einstein's special and general theories of relativity.
We think of advances in science proceeding at a dizzying rate, but
radical rearrangements in our outlook come slowly. Despite all our
technological advances, that is as true today as it ever was.

39   Cosmology's New Age

WMAP has started to sort through the possibilities of what happened in the first trillionth of a trillionth of a second, ruling out
well-known textbook models for the first time.

WMAP report on early-universe models

The initial measurement of the CMBR transformed the study
of the early universe from a field dominated by theoretical speculations to one in which precise measurements were abundant. After the
initial detection of the radiation, the most important question was to
confirm if what was being seen was indeed due to photons having
been freed from their interactions with surrounding matter four hundred thousand years after the Big Bang.

It seemed likely, but proof was needed and this required demonstrating that the photons had then been in thermal equilibrium. A
century of studies in thermodynamics and statistical mechanics specified exactly what was expected; saying that a room is in thermal equilibrium at a given temperature does not mean that every air molecule
or even every oxygen or nitrogen molecule in the room has the same
energy, but rather that their energies have a precisely specified distribution about a median. Temperature is nothing more than a con-
venient label for the peak in that distribution of molecule number
versus molecule energy.

In line with this reasoning, if the primordial photons were undisturbed remnants of a state in thermal equilibrium, their energies should reflect their original distribution. Penzias and Wilson's measurements had been taken at a single wavelength; since photon energy is simply related to photon wavelength, the goal was to determine the full range of radiation intensity versus wavelength and to check that the distribution followed the expected curve. One point was not enough to convince skeptics that the whole three-degree curve was there.

The Princeton group went ahead with their planned measurement at 3.2 centimeters. A sigh of relief was heard when the results were found to be consistent with expectations. Other groups, excited by the prospect of entering this new and dynamic field of cosmology, set out to make their own determinations. Within a few years seven additional points, at wavelengths ranging from 75 centimeters to 3 millimeters, were available. The time for celebrating had come: Arno Penzias and Robert Wilson shared the 1978 Nobel Prize in Physics *"for their discovery of the cosmic microwave background radiation."* Penzias's Nobel Prize acceptance speech, entitled "The Formation of the Elements," provided a thorough discussion of Geo's role.

The story wasn't quite finished, though. The curve's full extent needed to be traced out in order to be absolutely sure that what was seen corresponded to a once-existent thermal equilibrium. The expected curve of radiation intensity versus wavelength rises from the microwave region and attains a peak near a wavelength of two millimeters, or, equivalently, 5 waves per centimeter; it then falls precipitously. But the earth's atmospheric opacity made it impossible to confirm the predicted spectrum near and beyond the peak. It was quickly realized that only a satellite could provide definitive results. Accordingly, NASA began in the mid-1970s making plans for a mission dedicated to this goal, but a number of delays postponed the launch until mid-November 1989. In the meantime, a 1987 study from a rocket launched to an altitude of two hundred miles had obtained results that were clearly inconsistent with expectations. The measurements were taken at three different wavelengths, and in each case the answer was strikingly different from the expected, well beyond margins of error. Moreover, earlier balloon flights, though not as conclusive, had reported similar results.

The matter needed to be settled. In January 1990, only two months after the satellite was launched, John Mather, the overall leader of the NASA project, known by its code name COBE (Cosmic Background Explorer), was ready to present results. Sent to an altitude of almost six hundred miles, COBE carried three instruments, each designed for a special purpose. Mather's particular charge, FIRAS (Far Infrared Absolute Spectrophotometer), was responsible for tracing the full intensity versus wavelength curve. As Mather stood up to make his presentation to a packed audience at the annual meeting of the American Astronomical Society, the tension was palpable. The results would have been eagerly awaited in any case, but the earlier data from the balloon experiments had heightened the anticipation.

Mather's data, as shown in figure 6, had no ambiguity whatsoever.

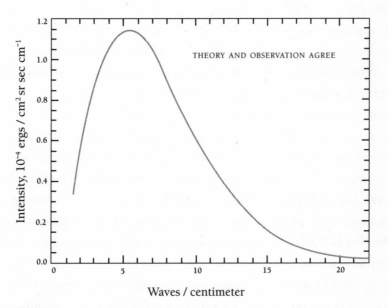

FIGURE 6
Cosmic Microwave Background Spectrum from COBE

It fit to within 1 percent the expected curve for radiation in thermal equilibrium. The audience stood up to give Mather and the COBE team a standing ovation. Some present that day have described the occasion to me as the most moving event of their scientific lives. The universe was now known with essentially absolute certainty to have once been in thermal equilibrium at close to three thousand degrees. Moreover, it was now clear that there had once been what Fred Hoyle had derisively called a Big Bang.

COBE's first report was spectacular, but there was more to come, as the satellite had three instruments on board and Mather had reported on only the results from FIRAS, the one dedicated to confirming or refuting the radiation's existence. After his report, the most eagerly anticipated new results were those from the third instrument, DMR (Differential Microwave Radiometer). Its purpose was to compare radiation at many points in the sky, each separated from the others by a fixed angle, in order to see if small differences appeared when moving from point to point. Although, as I've said, the relic radiation is expected to reach us uniformly from all directions, kernels of diversity needed to be in place or the uniformity would have lasted to this day. Even the greatest of trees once started with a small seed.

Estimates of these so-called anisotropies using models of what the universe was like in the wake of the Big Bang pointed to deviations from uniformity of approximately one part per one hundred thousand in the measured temperature. Detecting those deviations presented a formidable challenge, but it was within reach of the DMR instrument's capabilities. It took two years to complete the analysis. On April 23, 1992, George Smoot, the leader of the DMR team, made his presentation to the expectant audience of the American Physical Society in Washington. Agreement between the observed data and the theoretical predictions was striking. It was once again time to celebrate.

As mentioned, Arno Penzias and Robert Wilson shared the 1978 Nobel Prize in Physics *"for their discovery of the cosmic microwave background radiation."* John Mather and George Smoot divided the

2006 prize *"for their discovery of the blackbody form and anisotropy of the cosmic microwave background radiation."* The first had been bestowed for ushering in modern experimental cosmology. The second was for demonstrating in exquisite detail what the universe was like a few hundred thousand years after the Big Bang.

Nor did the march stop with COBE. A field that began with a handful of theorists and two individuals working on a small antenna in New Jersey is now one of science's main research areas. Hundreds, if not thousands, of researchers worldwide have joined forces in an intricate web of collaboration. Telescopes, balloons, and satellites bring specialized instruments high above the atmosphere. I hear the buzz— a different kind of buzz, of course, from what Penzias and Wilson detected—in the halls of any major physics or astronomy department. My own colleagues talk excitedly of the latest results they gather from the Atacama Cosmology Telescope, located at seventeen thousand feet in the high Chilean plains, and of the thrill they get recovering data from BLAST (Balloon-borne Large Aperture Sub-millimeter Telescope), in flight above Antarctica, even as they plan for future experiments.

MAP (Microwave Anisotropy Probe), a satellite launched by NASA in June 2001, has carried out the most comprehensive study of the sky. In 2003 it was renamed WMAP, the Wilkinson Microwave Anisotropy Probe, the prefix added to honor David Wilkinson, who died the previous year. He was the same David Wilkinson who, as a young Princeton physics instructor in 1963, set out with Dicke and Roll to build an antenna to look for that mysterious glow of the early universe. His forty-year run in cosmology, from 1962 to 2002, is something that neither he nor anyone else could possibly have imagined. It began in a university laboratory with three friends building a small antenna and ended with the gathering of a large international group to study the results from a satellite placed in orbit a million miles away from the earth. In both cases, Wilkinson was looking for signals emitted in the wake of the Big Bang.

WMAP's success was immediately apparent; its reports are the three most cited publications in all of physics and astronomy during the first decade of this century, and in 2003 *Science* magazine chose the probe's findings as the "Breakthrough of the Year." The mission was described as having achieved a consensus in which *"All the arguments of the last few decades about the basic properties of the universe— its age, its expansion rate, its composition, its density—have all been settled in one fell swoop."*

Instruments such as WMAP have made cosmology enter what many call its golden age, but the excitement in the field is due as much to the magnitude of the remaining problems as to the new instruments. It often happens in science that solving one problem reveals the presence of even bigger ones, and cosmology is a case in point. *Science* magazine said that the arguments of the last few decades had all been settled, but what will be those of the next few? One is clearly centered on the universe's composition. WMAP has announced, with a precision of approximately 1.0 percent, that atoms make up 4.6 percent of our almost fourteen-billion-year-old universe's contents. The remainder is distributed into dark matter, 23.3 percent, and dark energy, 72.1 percent. Since the identity of both dark matter and dark energy is not known, problems will continue to abound for future generations. In discovering more and more about our universe, we have unexpectedly found that we know less and less about what it is made of, surely a great science puzzle for our future.

To appreciate how much our perspective has changed, think back to what difficulty scientists had fifty years ago in even considering the early universe. Compare that trepidation to the relative comfort a contemporary scientist feels in reading WMAP's official Web page entry: *"WMAP has started to sort through the possibilities of what happened in the first trillionth of a trillionth of a second, ruling out well-known textbook models for the first time."* One cannot help being struck by that *"trillionth of a trillionth,"* but I also find the use of *"well-known"* striking. We have come to a point where such speculation is accepted, reasonable, and even experimentally tested.

40 Einstein's Biggest Blunder

Later, when I was discussing cosmological problems with Einstein,
he remarked that the introduction of the cosmological term was the
biggest blunder he ever made in his life.

Geo, on a conversation with Einstein

In 1933 Fritz Zwicky proposed the existence of what we now
call dark matter. Observations he had made of the Coma galaxy cluster indicated an abundance of a nonluminous substance dispersed
throughout its reach. This new material emitted no discernible signals,
making its presence known only by the gravitational force it exerted
on ordinary luminous matter. Seemingly far-fetched, though not more
so than his other proposal of neutron stars, Zwicky's suggestion was
not taken very seriously and lay dormant for the next forty years.

In the 1970s, Vera Rubin, Geo's former graduate student, took up
the question. Brilliantly performing a systematic analysis of numerous galaxies, she showed that the paths of stars inside them could be
explained only if a great deal of dark matter was present. Since then,
her studies have been extended, and we now have evidence for this
mysterious substance throughout the universe. Many searches for it
are presently being undertaken, but its identity remains an enigma; a
current favorite among answers is some novel type of massive elementary particles. But discovering them, which many physicists hope will

take place at the new accelerator known as the LHC (Large Hadron Collider), will not help explain the even greater puzzle of dark energy. All matter, dark or not, manifests itself by attracting its counterparts through the force of gravity; dark energy's action, on the other hand, is repulsive.

For many, the most likely interpretation for this strange appearance is the existence of a cosmological constant, the term Einstein once inserted into his equations of general relativity and later discarded, famously telling Geo that its introduction was *"the biggest blunder he ever made in his life."*

When Einstein's general relativity equations are considered for the universe as a whole, as Einstein was quick to note in 1917, they imply that the universe's rate of deceleration of expansion is proportional to its average energy density, calculated by taking into account stars, planets, dust, and whatever else, seen or unseen, exerts a gravitational force. The two sides of Einstein's equations have a very different nature; one depends only on very general assumptions, while the other is a moving target, changing as our knowledge of the average energy density evolves. Think of the one as marmoreal, perfectly shaped by the master sculptor, and the other as formed from soft clay. The latter is subject to the vagaries of our knowledge of the surrounding cosmos and, most mysteriously, to the possible presence of what we have come to call the cosmological constant, a measure of the energy density of the vacuum.

You might object, claiming that the vacuum is really nothing more than the void, the absence of everything, but there is a subtlety. Might it be that we can know only increments, relative changes to an otherwise constant background that we have chosen to call the vacuum? Such an apparently strange notion has in fact become a central theme in twentieth-century physics. When in 1928 Paul Dirac joined quantum mechanics to special relativity, his synthesis faced a perplexing prediction, particles with negative energy. Since these were manifestly nonexistent, another way of overcoming the obstacle had to be found. The eventual solution involved the existence of antimatter, but the first step along that path of discovery required a reinterpretation of the vacuum, which forced physicists to reexamine what apparent nothingness really meant.

Such a notion, though prima facie surprising, has existed in one form or another since the Pythagoreans proposed it 2,500 years ago. Their universe consisted of the stars they saw in the night sky, the earth, the moon, the sun, and the planets. They thought these were all moving spheres suspended in space, all completing circular orbits about a central fire that was hidden from the earth. Hoping to discover a rationale for the planets' distance from the sun, they sought guidance in their central credo, *"All things are numbers."* Since the original Pythagorean belief in the paramount importance of numbers had been strengthened by their discovery that notes emitted by plucked strings are harmonious only if the string lengths are fixed as simple ratios, they looked to the combination of numbers and music for a clue. That is how, in a leap that joined two seemingly unconnected areas, they arrived at a notion that answered their query. Each planet supposedly emitted a musical note as it marched along on its path, the note's frequency set by its distance from the central fire. The ensuing ensemble would be a melody that filled our world.

The Pythagoreans reasoned that just as the sounds from plucked strings are harmonious for only certain ratios of string length, the notes emitted by planets would be pleasing only if the lengths of the planets' radii stood to one another in a relation analogous to that of the plucked strings. Why were we not aware of this celestial song? The Pythagorean response was that the melody existed as a constant background against which our lives played out. Aware only of changes, we had not realized its constant presence even though we had been hearing the tune since birth.

Plato and Aristotle considered this notion, as did others over the next two thousand years. Capturing the imagination, the image of melodious planets was enshrined in art and literature. Elizabethan audiences were not puzzled by Lorenzo's reference to these inaudible sounds in *The Merchant of Venice:*

> *There's not the smallest orb which thou behold'st*
> *But in his motion like an angel sings,*
> *Still choiring to the young-eyed cherubins.*
> *Such harmony is in immortal souls,*
> *But whilst this muddy vesture of decay*
> *Doth grossly close it in, we cannot hear it.*

But the analogy between Pythagorean thought and modern cosmology is not perfect, for we *can* hear the cosmological constant, though not by tracing the paths of falling apples or even those of planetary orbits. In order to do so, we must study the relative expansion or contraction of the universe. In 1917, laboring under the then-prevalent belief in a static universe, Einstein introduced the constant into his equations to have them agree with his preconception. A dozen years later, when Hubble showed that the universe was in fact expanding, Einstein did away with the constant, eventually making his famous remark that introducing the constant had been his *"biggest blunder."* Now, more than fifty years after his death, the possible need for such a term has resurfaced, though neither for the reason Einstein originally desired it nor with the magnitude he selected.

When we started to think that the universe began in a Big Bang and that there was no cosmological constant, we asked whether the universe would continue expanding or, at some point, reverse its motion. Would a Big Crunch follow the Big Bang? The situation was schematically viewed as similar to a rocket shot up in the air: Does it forever recede from us, decelerating as it gives up the energy needed to overcome the earth's gravitational attraction, or does it expend all the energy and fall back to earth? The question was not if the rate of the universe's expansion decreased, but by how much.

This was the state of affairs until a little over a decade ago. At that time the astronomy and physics communities were shaken by the precision measurements made by two California-based groups. They were looking at very faraway type-1a supernovae. These are stellar explosions that emit known quantities of light, making them "standard candles," like the much closer Cepheid variable stars Henrietta Leavitt studied. Given their great distance from earth, the blasts necessarily occurred billions of years ago, allowing the two teams to analyze calibrated "candles" farther away than any previously known ones.

Though both groups were based in Berkeley, they came from different scientific communities. The Supernova Cosmology Project originated largely as a team of experimental physicists with expertise in analyzing large sets of data. The High-Z Supernova Search Team consisted of astronomers with a special interest in supernovae. Though their styles were different and a certain amount of rivalry was present, both groups reached the same startling conclusion almost simultaneously: the universe's rate of expansion was increasing, a result since then confirmed by other groups, including WMAP.

A new kind of antigravity, or repulsion, appears to be overcoming the gravitational attraction felt by the universe's matter. For want of a better word, this mysterious presence has come to be known as dark energy. A cosmological constant is the prime candidate for this mysterious presence, and elucidating its nature or obtaining a satisfactory alternative explanation for the universe's observed accelerating expansion is the greatest problem in present-day cosmology.

Geo would have loved this situation: a strange form of antigravity, perhaps determined by an unexplained but crucially important number seemingly unconnected to any known theoretical physics predictions. This is begging for one of his "crazy ideas." Perhaps, as homage to Geo and his love of both limericks and *Alice's Adventures in Wonderland*, I should summarize the situation by attempting one of my own:

Physicists are happy to chatter
About the existence of dark matter

But when a constant appears
It confirms their hidden fears
That the universe is as mad as a Hatter.

Moreover, if the cosmological constant is indeed the reason for the universe's accelerating expansion, its dominance will grow over time. The reason is that matter becomes increasingly less dense as the universe's volume increases. A constant, on the other hand, is the same now as in the past or the future. This means that in the distant past, when the universe was much smaller, the cosmological constant was essentially irrelevant. But in a future time, when the universe is a hundred times larger in scale or a million times larger in volume, it is matter density that will be irrelevant, a million times smaller than the cosmological constant's contribution.

Why do we happen to be living in that phase of the universe's life when the abundance of matter and the constant are comparable? This is the kind of question that keeps physicists awake at night.

41 Duckling or Swan?

Well, the old horse is dead, so why kick it again. But I was very unhappy because Dirac's idea is so elegant. Thus 48 hours ago I told myself . . .

Geo, on how an idea came to him

Some physicists have begun considering that Einstein's equations of general relativity may have to be modified even beyond the introduction of a cosmological constant. I described these equations as a relation between two quantities, one perfect and immutable and the other subject to the vagaries of our measurements of the universe's energy content. But even the "perfect and immutable" part was the result of a choice made by Einstein; it seems inevitable, because altering it would violate the simplicity tenets of Occam's razor, as well as our criteria of beauty and elegance. But one generation's choices of simplicity, beauty, and elegance may seem unnatural, awkward, and misguided to its heirs. As a famous Russian physicist once said, *"The garbage of the past often becomes the treasure of the present"* (and vice versa).

I have often heard colleagues announce that a model they devised must be right because it is so beautiful, only for them to be quickly disappointed. On the other hand, creations that seem awkward and clumsy may yet turn out to be beautiful when viewed from another perspective, just as the ugly duckling of the childhood fable altered its perception of itself when it realized it was a swan. Perhaps the cosmological constant will suffer a similar fate. Our present theories have no explanation for its observed value, but future generations may see it as a graceful, long-necked swan.

Many physicists think the root of the problem lies in our not having yet obtained a successful unification of general relativity with quantum mechanics. They are considered the two greatest achievements of twentieth-century theoretical physics, but joining them together has proved elusive. Superstring theory has presented itself as a formidable candidate for the union, but so far it has not received universal acceptance, nor has it yet reached the point where it is able to make unambiguous, testable predictions. In the meantime, the calculation of the cosmological constant, or whatever else acts in its place to effect the universe's accelerating expansion, continues to present a puzzle for superstring theory and any other candidate for the marriage of the two great theories.

Before the discovery of dark energy, the prevailing belief held that we were missing some general symmetry requirement that would have to be imposed on our theories, one that would automatically require this curious parameter we call the cosmological constant to vanish. Candidates for how to achieve the desired goal, none of them completely convincing, were offered, but there was hope that some as-yet-undiscovered principle would do the job. Now we are faced with the dual challenge of finding why the constant is there at all and, yet, is so much smaller than expected.

"Smaller than expected" is a questionable concept if you don't really have an accepted theory with which to calculate a response. In this case the use of the word *smaller* is meant to reflect the conclusion that simple estimates of the cosmological constant yield answers that are considerably larger than the observed value. This is of course little more than an indication that we lack some key ingredient in making the estimate, but the discrepancy between calculations and observation is striking. As an example, one scheme that has attracted considerable attention yields a value larger than what is observed by a factor of approximately 10 to the 39th power cubed, or 10 followed by 116 zeroes.

I quote this value in part because the appearance of 10 to the 39th power would have reawakened one of Geo's long-standing interests. In 1937 Paul Dirac published a short paper in *Nature* in which he pointed out that the ratio of the electric and gravitational forces between an electron and a proton is approximately 10 to the 39th, and noted that this was also the age of the universe, when measured in time units that seemed natural to him. Believing that the equality of the universe's age and the ratio of forces was no accident, he thought it must continue to hold even as the universe aged. In other words, the ratio was increasing.

Coming from anybody else, such conjectures would have appeared as nothing more than the observation of a curious coincidence. However, the effect of the pronouncement by Dirac was startling, because most physicists regarded him as second only to Einstein in originality and power of insight. One need only look at Max's 1978 Caltech oral interview to gauge this opinion. When asked if Dirac, four years his senior, was a friend, Max's reply was:

> *That would be an exaggeration. He was my hero. I mean I had an infinite admiration for him, and studied every one of his papers and his book when it came out, but I was far too much in awe of him to be close.*

Up until the appearance of the paper featuring 10 to the 39th, Dirac had been known for extraordinary powers of logic and an adherence to rigorous precepts, but this was different. Nor was it the only shock he delivered in 1937. On the occasion of being awarded the 1933 Nobel Prize in Physics (shared with Schrödinger), Dirac had been described by the London newspaper reporter interviewing him as being *"shy as a gazelle and as modest as a Victorian maid."* But in 1937, to the astonishment of those who knew his retiring ways, Dirac married. Was a new Dirac emerging? It seemed hard to believe. The succeeding years would reassure most physicists, revealing that his way of thinking had not altered, and this conjecture was more an anomaly than a reflection of a changed Dirac.

Old or new Dirac, Geo found his "10 to the 39th" suggestion entrancing, but others seemed to take a more jaundiced view. Geo recollected how the notion had struck Bohr when he encountered it:

> *The first criticism of this idea was made by Bohr. I still remember him coming into my room (I was visiting Copenhagen at that time) with the fresh issue of "Nature" in his hands, saying "look what happens to people when they get married."*

Bohr's joking reference was of course to how unusual it was to see a conjecture of this sort coming from Dirac.

Over the years Geo continued to think about Dirac's suggestion. Others, principally Geo's old friend Edward Teller, thought about it as well. In 1948 Teller showed that a combination of geological records and the accepted model of solar production ruled out the notion that the gravitational force grew weaker over time in the way Dirac envisioned. Gamow refined Teller's argument in 1967 and, in doing so, began to reconsider Dirac's idea. Geo then had an idea of his own. In a letter he sent to his old friend Phil Abelson, the editor of *Science,* he detailed how it came to him. He had submitted an article to *Science* two days earlier, and now, writing from a Denver hospital bed where he was convalescing after having calcium deposits removed from his carotid arteries, Geo told Abelson, *"Well, the increased supply of fresh blood to my brain caused a brain storm (fortunately not a brain stroke) and the very evening of the day that I mailed that letter I got an idea."*

The original conjecture that the gravitational force was decreasing over time might have been disproved, but Geo was not ready to give up. *"Well, the old horse is dead, so why kick it again. But I was very unhappy because Dirac's idea is so elegant. Thus 48 hours ago I told myself . . ."*

Geo now suggested that Dirac's conjecture of an increasing ratio of electric to gravitational forces could still work if instead of gravitational forces decreasing, electric forces increased in strength. But Freeman Dyson quickly and brilliantly showed that this type of variation could also be ruled out. He did so in months rather than the years it had taken to disprove Dirac's original version of the conjecture. It seemed once again that the apparent equality of force ratio and universe age was simply a coincidence.

42   After the Golden Age

Had anyone suggested in 1953 that the entire human genome would
be sequenced within fifty years, Crick and I would have laughed and
bought them another drink.

James Watson, writing in 2003

Cosmology's dark ages persisted for approximately half a century, assuming we consider modern cosmology to have begun with
Einstein's general theory of relativity. It was still possible in the early
1960s to think of the universe as having no beginning and no end or
as experiencing a fiery origin in a Big Bang. After 1965, holding on to
a belief in the steady state model's eternal universe became increasingly problematic.

Similarly, although the modern study of genetics began around
1900, it is fair to call the era before 1953 the subject's dark ages,
despite the many brilliant insights of the preceding years. They were
dark in the sense that the identity of the gene remained shrouded in
mystery. To be sure, Beadle and Tatum's famous *"one gene, one enzyme"*
dictum was a clear indicator of how genes act, and Avery's experiments pointed to DNA as the carrier of genetic information, but until
1953, it was still perfectly arguable that genes were protein agglomerates. After the Watson-Crick model, this was no longer acceptable.
Genetics now entered what some call its classical phase, one in which
Crick's central dogma that *"DNA makes RNA makes protein"* was laid

down, the genetic code was deciphered, and the RNA agents necessary for the workings of the progression had their existence confirmed experimentally. A third stage in genetics then began. Great discoveries have been and continue to be made in this new phase, but they have tended to be incremental rather than paradigm shifting.

This is a common pattern in science. Sometimes an advance in a theory points the way to the acceptance of an emerging picture, and sometimes an experimental result does so, but the eventual survival of the whole hinges on their coming together. Cosmology is no exception. However, the shift in thinking was less radical than it was in genetics, because the theory of the Big Bang had laid the groundwork for interpreting the experimental discovery of that cosmic microwave background radiation, whereas nobody had been prepared for the simplicity of DNA's double helix.

It is also true that Geo's role in cosmology's golden age, the post-1965 era, was negligible, just as Max's was in genetics. But both are viewed as pioneers who sensed the correct direction for their disciplines, made crucial contributions, and prodded others to follow them. They led the way. A great deal of work remained to be done in each subject, but the cornerstones had been laid.

The thirteen years between Watson and Crick's 1953 discovery of the structure of DNA and Max's 1966 Festschrift were extraordinarily exciting. By that period's end, the way in which DNA uncoiled was no longer a mystery, and the various forms of RNA that constitute the links between DNA and amino acids were known. Surprises had appeared and been dealt with. Many think none was as remarkable as the appearance of the regulatory genes that serve to activate or inactivate other genes. Conjectured by François Jacob and Jacques Monod, these genes function by producing proteins (repressors) that block other genes from being read by messenger RNA when the enzymes those genes produce are not needed. It seemed to be a baroque construction, but evolution, as we have already noticed, does not always follow the precepts of Occam's razor. Some thought the repressor scheme was simply too complicated to be true, but all doubts about the regulatory genes' existence vanished in 1966, the same year as the Festschrift. Two Harvard teams, operating only a corridor apart, managed almost simultaneously to find repressors in completely different organisms. Walter Gilbert and his postdoctoral fellow Benno Müller-Hill discovered them in a bacterial system, and their neighbor Mark Ptashne found them in the viruses that infect bacteria.

Observing the field from a distance in those years, I couldn't help feeling envious. Walter Gilbert, commonly known as Wally, had been my faculty adviser in 1959, when I was a college senior. At that time he was a twenty-seven-year-old up-and-coming assistant professor in the Harvard physics department, an elementary particle theoretical physicist engaged in mathematically oriented research. He seemed to be succeeding in a way I could only dream of. A few years later, I noted that Wally had switched to molecular biology; he was now doing pioneering work as an experimentalist in a totally different field. As for Mark Ptashne, we are the same age and had been friends as graduate students. I saw somewhat wistfully that he had become wildly successful while I was still looking for an assistant professorship. Had I made a mistake in my selection of a career?

After 1966 a new generation sprang up in molecular biology, one prepared to fully exploit the possibilities offered by understanding an organism's DNA and RNA. By the early 1970s techniques had been developed for cutting DNA at desired sites along the chain, inserting an extracted piece of DNA into a bacterial cell, and, thanks to the speed with which bacteria reproduce, rapidly making multiple copies of the segment. DNA cloning was born.

The next important tool to be developed was a means of reading the piece of code incorporated on those many copies of the DNA segment, or, in the common language used to describe the process, sequencing them. Two methods were developed independently within a few years. Wally Gilbert found one; Fred Sanger in Cambridge, England, developed the other. Sanger had already won the 1958 Nobel Prize in

Chemistry for determining the sequence of amino acids that constitute insulin, the first complete reconstruction of a protein. He was now awarded the 1980 prize, his second, together with Gilbert *"for their contributions concerning the determination of base sequences in nucleic acids."*

With instruments in hand and the assistance of the massive computing systems that had been developed in the interim, scientists began attempting to sequence the genetic code of different organisms. The human genome was the ultimate goal; finding the identity of its 3.1 billion base pairs would be to genetics what ascending Mount Everest had been to mountaineering, certainly not the end of the adventure and not even the most difficult feat ever attempted, but a goal that everyone would recognize as a turning point.

Watson wrote in 2003, *"Had anyone suggested in 1953 that the entire human genome would be sequenced within fifty years, Crick and I would have laughed and bought them another drink."* By 1980 what seemed laughable in 1953 was becoming a very real possibility. Serious discussions on sequencing the human genome began with none other than Watson leading the charge. The preliminary estimate was that it would cost three billion dollars, an expense dwarfing in magnitude that for any science project except space missions and the very largest high-energy physics particle accelerators. The sheer size of the undertaking led many in the biology community to fear it would drain funding from all other genetics research; admittedly it would be a giant technology effort, and very useful as such, but might it be that science would be advanced further by funding myriad smaller projects? These objections were countered by the claim that a project of this magnitude would promote public understanding and galvanize support for genetics in a way nothing else could. It would do for molecular biology what the moon landing had done for space research.

With some trepidation, the Human Genome Project officially began in October 1990. On June 26, 2000, U.S. president Bill Clinton and UK prime minister Tony Blair made a simultaneous announcement that it had been completed. This was, in fact, an overstatement; only a rough draft had been finished, but within the next three years the goal of reading close to 95 percent of the genome with an error rate of less than one in ten thousand bases had been reached.

The first genome had been assembled as a composite picture from the DNA of many donors, so a new milestone of sorts was reached with the complete decoding of an individual's DNA. Symbolically, Jim Watson was chosen for this initial venture. The task was completed at a cost of one million dollars; in May 2007, he was presented with two DVDs containing a complete copy of his genome. In a gesture that is more than simply symbolic, for it may help in medical research, Watson immediately made the sequence available online. I probably should say that an *almost complete* copy of his genome was made available to him and to the public: illustrating some of the possible negatives that may come with this wealth of knowledge, Watson asked to not be informed of the status of his apolipoprotein E gene, believed to be a risk factor for Alzheimer's disease.

The point where our genome becomes a routine part of our medical record is not that far in the future. Though the initial genome sequence cost billions of dollars, the expense in 2007 for Watson's was only one million dollars, and the price is continuing to drop. In 2010 it stands at fifty thousand dollars, and within a few years technology will have almost certainly been developed to read the three-billion-letter book at a cost of a thousand dollars, the commonly accepted marker for having human genome sequencing become a commonplace event. All of us will then have to ask ourselves what we want to know of our genetic record. We saw the decision Watson made. Most of us would like to be informed of an impending disaster that can be averted, but how many of us wish to know of one that cannot?

Studying the genomes of organisms, a field known as genomics, is advancing at a staggering pace. I have weekly conversations with my daughter Julie, who directs a laboratory at the National Institutes of Health's National Human Genome Research Institute. We might discuss the recent announcement of "synthetic life" production made by the ever-controversial Craig Venter, the endeavors by Eric Lander and the Broad Institute he heads to use genomics to study mammalian disease, or Julie's own work on the Human Microbiome Project, an effort to sequence the bacteria inhabiting the human body; there are over a thousand species in the human gut alone.

Max and Geo would have been impressed.

43 The Unavoidable and the Unfashionable

Space, cosmology, origin of life, futurology, I find it all chilling. I always admired Gamow who found these things "fun" of which I was scared. Admired? Wrong word. Amazed, but also shocked by the frivolity of it.

Max, writing to Freeman Dyson in 1970

The new generation of genomics researchers lives in a technology-driven, interconnected world that might have seemed alien to the small band forming the molecular biology world a half century ago. Some of those pioneers embraced it, but others were already beginning to look in alternative directions by the 1960s. This trend can be seen in Max's 1966 Festschrift. In the introduction to the published volume, Gunther Stent says, *"Now that the success of molecular genetics has made it an academic discipline, one can expect that in the coming years students of the nervous system, rather than geneticists, will form the avant-garde of biological research."*

And in the years after that Festschrift, many of molecular biology's early trailblazers did in fact turn to neurobiology; Stent himself began to study the leech's nervous system. But nobody made a more flamboyant transition than Francis Crick.

In 1976 he left Cambridge, England, for California's Salk Institute, a place he had been visiting for more than fifteen years. In making the move, Crick opted to leapfrog simpler systems and begin straightaway by studying the human brain, an organ with approximately one hundred billion nerve cells (neurons). Twenty years later he summarized his views in a book for the general public, *The Astonishing Hypothesis,* stating his premise on page one:

> *You, your joys and your sorrows, your memories and your ambitions, your personal identity and free will, are in fact no more than the behavior of a vast assembly of nerve cells and their associated molecules.*

It is a book, as he acknowledges, about how to *"come to grips with the problem of consciousness"* and what sort of experiments need to be carried out in order to understand in scientific terms this very essence of what it means to be human, a problem that had always interested Max.

Luria once compared Max to Crick. When asked if his old friend was good at picking people, he replied:

> *It's not so much that he is good at picking people, as that he is attracting to people. Because he is terribly intelligent. Because it is so very exciting to work with him. His ideas, the way he thinks, the order—I find Francis Crick, for example, probably the only other person who is equally, who is exciting in that same way.*

Not surprisingly, given the assessment Luria made, Max had been entertaining thoughts similar to the ones propelling Crick into this new field. In 1974, two years before Crick's move to the Salk Institute, Max delivered a set of twenty lectures at Caltech and repeated them in a modified version in 1976. They were assembled in a single volume entitled *Mind from Matter? An Essay on Evolutionary Epistemology.* The tone is set on page one, where Max asks *"what do we know and how do we know it?"*

An enigmatic quote from an 1846 diary entry by the Danish philosopher Søren Kierkegaard prefaces the lectures:

> That a man should simply and profoundly say that he cannot understand how consciousness comes into existence—is perfectly natural. But that a man should glue his eye to a microscope and stare and stare and stare—and still not be able to see how it happens is ridiculous.

Max's response to that challenge is to say that we need

> to do that ridiculous thing, "look through the microscope," to try to understand how consciousness or, more generally, how mind came into existence. And with mind, how language, the notion of truth, logic, mathematics, and the sciences came into the world. Ridiculous or not, to look for the evolutionary origin of mind today is no longer an idle speculation. It has become an approachable—indeed an unavoidable—question.

But what exactly did Max mean? Like Bohr, he loved a paradoxical turn of phrase and could be cryptic even when he seemed to be clearest.

Crick was asking how we *"come to grips with the problem of consciousness,"* and Max was trying *"to understand how consciousness or, more generally, how mind came into existence."* I strongly suspect that Max was also wondering if some new laws of physics might eventually be needed to comprehend these phenomena, just as Bohr and Schrödinger had questioned the need for such laws in understanding the basics of life. He also certainly questioned, as Stent says in Max's Festschrift, whether the brain,

> being a finite engine, may not be capable, in the last analysis, of providing an explanation for itself. In that case the paradox will have

been found at last: there exist processes which, though they clearly obey the laws of physics, can never be understood.

In the ensuing years, efforts to invoke new laws of physics to understand the human brain have largely been dismissed, just as the need for such new laws to understand the basics of life were dispelled by the appearance of the double helix model of DNA. But we have come to realize that there is much more to understanding a creature's existence than knowing its genome. Likewise, further notions of complexity will certainly be needed to describe the human brain, an organ with a hundred billion neurons and a thousand times that many connections. Even assuming the computer analogy is a good first step in that direction, we can ask if it is a digital or an analog device. Insofar as comparisons are fair, the similarity in the height of the nerve impulses traveling along the brain's nerve fiber connections suggests digital transmission, but the information carried by the varied timing of those impulses points to analog. Once again the most likely response is *"Evolution is cleverer than you are."* Being so, it almost certainly made use of all the tricks in its grab bag, including some we are probably not yet aware of.

However, even if we accept that evolution is so very clever, it still seems remarkable that we have managed to acquire, in Max's words, *"language, the notion of truth, logic, mathematics, and the sciences"* in the few million years since our species split off from our common ancestors. How have we done so? It is not a matter of massive genetic differences, given that we share 99 percent of our genome with chimpanzees; so what is it? I was thinking about this question and how it might relate to Max's interests during a recent visit I paid to Freeman Dyson at the Institute for Advanced Study in Princeton.

Dyson has a well-earned reputation for brilliance, breadth of views, and ability to extract interesting lessons from gazing into science's future. He has written extensively and eloquently on a great number of such questions. While many do not agree with the conclusions he has drawn, this does not appear to bother Dyson; he likes to challenge conventional views. When I asked him for his thoughts on the question of how humans had developed their special skills, Dyson told me about his interest in recent research regarding mutational differences in a section of DNA that regulates gene behavior involved in the development of the cerebral cortex.

Dyson thought these developments were very exciting, and I am sure Max would have as well, but I also suspect Max would have found them too *"fashionable"* and would have looked for some out-of-the-way, seemingly obscure topic for his own research. Consulting the eminent neuroscientists Lily and Yuh Nung Jan, a wife-and-husband team at the University of California at San Francisco, reinforced my view on Max's preferences. They came to Caltech from Taiwan at the end of the 1960s to study theoretical physics but, under Max's influence, soon shifted to neuroscience. Yuh Nung wrote a Ph.D. dissertation under Max's direction and then took a postdoctoral fellowship at Caltech with Seymour Benzer. Remembering his two mentors, Yuh Nung Jan commented on how *"Both strived to do pioneering work on fundamentally important work and, at the same time, stayed away from the crowd."* The Jan laboratory Web site is quite simple. Underneath a paragraph about their work, centered and in boldface type, it reads, *" 'Don't do fashionable science'—Max Delbrück."*

How would Max have squared this dictum with the equally emphatic *"to look for the evolutionary origin of mind today is no longer an idle speculation. It has become an approachable—indeed an unavoidable—question."* In other words, how would he have reconciled the need to do the unavoidable with his penchant for the unfashionable? He might have been torn, but in the end I think that the unfashionable would have prevailed. Max would have looked for some obscure organism that displayed curious behaviors, perhaps hoping to encounter a new Luria, someone who one day would say something akin to *"when Delbrück and I first met, we were probably the only two people interested in phage from the point of view of 'molecular biology.'"*

Dyson also told me, before my visit, about having corresponded with Max during the 1970s on a variety of questions, and he very kindly invited me to view their exchange. I hoped it would shed some light on Max's approach to science, and was not disappointed. One letter even contains a fascinating statement Max made about Geo and, by reflection, himself. On November 21, 1970, Max wrote Dyson:

Space, cosmology, origin of life, futurology—I find it all chilling. I always admired Gamow who found these things "fun" of which I was scared. Admired? Wrong word. Amazed, but also shocked by the frivolity of it. That's it: shocked by the brash frivolity, like that of our Arms Race Geniuses who will blow us all to pieces in short order. But then, they are not really that much different from the rest of us.

Max, who had studied the Greek classics and frequently referred to them, next invoked the myth of Tithonus, the lover of Eos, the goddess of dawn. She asked Zeus to give Tithonus eternal life, but forgot to add eternal youth, so her lover gradually became more and more enfeebled. In a variation, pity is taken on the old man and he is transformed into a cicada; following this version, Max concludes by saying, "*The Gods gave us Eternal Life, but not Eternal Youth. Like Tithonus, now we (scientists) chatter away incessantly, like crickets, but have lost the first blush of love of seeing the world. Sounds, and is, Victorian.*"

On December 3, Dyson responded, letting Max know that he felt scientists could, and some did, preserve their excitement at studying the world:

In recent years I have found one group of scientists, the astronomers, who have not lost their love of seeing the world in its wild beauty. While the cricket-like chatter of theorists chirps interminably on, the observers retire to their mountaintops and see things as they really are. Last year I even watched the Crab Nebula Pulsar pulsing away through our little telescope on the Princeton campus.

The heavens and earth are still full of God's glory, if you have your eyes open.

I had lunch with Dyson after reading the letters. We discussed them briefly, and I was pleased to see how interested he was to hear that Max had originally been an astronomy student at Göttingen and that patiently studying the heavens at night was a habit he preserved into old age.

Many who accompanied Max on the camping trips into the California desert had noticed him carefully recording his observations of nighttime sightings in a special notebook. When one night his son Jonathan asked what he was doing, Max replied,

> *"I am charting the movement of the planets around the sun, from what can be seen with the naked eye." When I asked him what he had concluded from his observations, he answered me in the way he usually did, by asking me to study the sketches, and what did I think?*

Max was adopting the Socratic method for his teaching. As for his observations, he eventually approached the precision of Tycho Brahe, the last astronomer to rely solely on the naked eye as an observational tool. I like to think that he was also remembering his youthful traipsing through his parents' bedroom to view the stars over his old home in Grunewald.

Perhaps that was the true secret of both Max's and Geo's success in science. Even as they aged, despite the travails and setbacks, their ups and downs, each struggled to approach new problems with *"the first blush of love of seeing the world."*

44 Mr. Tompkins Arrives

When I do not have any ideas to work on, I write a book; when some fruitful idea for scientific pursuit comes, writing lags.

Geo, describing his writing of books

Max never wrote any books. He had intended to publish the lectures that became *Mind from Matter?* but his final illness intervened, and the book appeared only after his death in an edited form, assembled by a few of his close friends. By contrast, Geo's production of books was astonishing both in quantity and in character, for they are quite unlike those by anybody else. I am not speaking of the technical ones, for interesting as they may be, they are accessible only to an audience of physicists and do not differ markedly from those written by others. It is rather the ones intended for a general audience that display his idiosyncratic but deep knowledge of many fields of science, his humor, and his playfulness. I can attest that few physicists of my generation, at least in the Western world, did not read one or more of them—there are nineteen altogether—while in their teens and close the cover hoping they could one day participate in the world Geo was describing.

My own favorite, *One, Two, Three . . . Infinity,* first appeared in 1947. Beginning with a limerick about a young fellow from Trinity who took the square root of infinity, it is a three-hundred-page romp through science and mathematics, with chapters on number theory, probability, atomic and nuclear theory, relativity, chemistry, astronomy, cosmology, and biology. Delightfully illustrated by Geo, it is full of humorous stories and limericks, with references to Mickey Mouse as well as to Einstein, Rutherford, Archimedes, and Aristotle.

The book is dedicated to *"my son Igor, who wanted to be a cowboy."* Fourteen years later, when it was reissued, Geo concluded the new preface by responding to the numerous inquiries he had received about Igor's subsequent career. He noted that Igor had changed his mind and was now graduating from the University of Colorado with a major in biology, intending to pursue a career in genetics. Geo does not mention that Igor, as colorful as his parents, had already been by then a motorcycle courier for CBS News and a dancer in the National Ballet Company. I can also attest that Igor's desire to be a cowboy has not altogether disappeared, for he still rides in the Boulder, Colorado, canyons every day, rain, shine, or snow.

The prose in *One, Two, Three . . . Infinity* is light and breezy, but throughout the book one senses the workings of a curious and deeply informed mind. Its tone is set in the preface, which begins with a quote from Lewis Carroll's *Through the Looking-Glass:* "'The time has come,' the Walrus said, 'To talk of many things.'" It concludes by acknowledging the help of the eleven-year-old daughter of Princeton's famous mathematician John von Neumann.

> *Above all my thanks are due to my young friend Marina von Neumann, who claims she knows everything better than her famous father, except of course mathematics, which she says she knows only equally well. After she had read in manuscript some of the chapters of the book, and told me about numerous things in it which she could not understand, I finally decided that the book is not for children as I had originally intended it to be.*

I agree with Marina that the book is not for eleven-year-olds, but looking it over makes me want to be again a fifteen-year-old would-be scientist so that I could encounter for the first time prime numbers, curved space, nuclear fusion, spiral galaxies, genetic crossings, a wonderful little discussion about viruses that includes a description of Max's work, and, of course, Geo's view of how the universe began.

Geo never wanted to popularize science in the usual way. This book and all his subsequent ones have a tone of "Come have fun with me" rather than of "Let me explain this to you." He had found a voice with which to do this in 1937, when he created C. G. H. Tompkins, a *"little clerk of a big city bank."* Commonly known simply as Mr. Tompkins, he is a bank employee with an insatiable curiosity. Geo used him in a 1937 story called "A Toy Universe," in which Tompkins is shrunk, stretched, and speeded up in a series of adventures that illustrate the theory of relativity. Geo submitted the piece to *Harper's, The Atlantic Monthly, Coronet,* and other magazines of the time. Somewhat discouraged after receiving polite but firm rejections from all, he put it away.

The following summer, in 1938, he was having a drink with Charles Darwin, a physicist and a grandson of the famous biologist, at a conference on "New Ideas in Physics" when the discussion turned to the subject of popular science writing. Geo mentioned his "Toy Universe," and Darwin told him there was a man in Cambridge, England, C. P. Snow, who might be interested in the piece, adding that Snow, the editor of a magazine called *Discovery,* seemed to be on the lookout for good writing about science. When he returned to Washington, Gamow sent Snow "A Toy Universe." He soon received a telegram: *"Your article will be published next issue. Send more. Snow."* Geo did, and six months later the manager of the American branch of Cambridge University Press suggested that the Mr. Tompkins stories from *Discovery* be put together with additional material and published in book form.

Meanwhile, a second letter arrived almost simultaneously with the one from Cambridge University Press. It was from a senior editor at Viking who had heard of the excitement in nuclear physics during the previous years and knew of Geo's earlier Oxford textbook. He asked if Geo would be interested in writing a book addressed to an interested but nonexpert public. The rapid reply was again an enthusiastic yes, but in order to make the venture more interesting, and above all more fun, Geo thought, why limit it to nuclear physics? He could certainly explain the subject's basics and touch on the recently discovered fission; the book's pièce de résistance would be the exciting discoveries in nuclear astrophysics that had been discussed at the George Washington University conference in 1938. However, if the general topic was the sun, he could also discuss the measurements that ancient astronomers made, Galileo's observations of sunspots, their relation to the sun's magnetic field, and ever so much more. He could talk about the sun's composition and attempts to place our very own star within the framework of all the other stars. In addition, since Geo loved drawing cartoons, he knew who the illustrator would be.

First appearing in 1940 with the title *The Birth and Death of the Sun*, the book has been reprinted six times, appeared in numerous paperback editions, and been translated into more than a score of languages, ranging from Danish to Hindi. It established Geo's reputation as a writer and set him off on a second career. *The Biography of the Earth* appeared in 1941, *Mr. Tompkins Explores the Atom* in 1944, *Atomic Energy in Cosmic and Human Life* in 1947, and *Gravity* in 1962. The previously mentioned *The Creation of the Universe*, Geo's response to Hoyle's book, was published in 1952, and *Thirty Years That Shook Physics* in 1966.

In 1964 Geo wrote a book titled *A Star Called the Sun*. Too many years had passed since *The Birth and Death of the Sun* for a book in an active research field to remain topical, so this was an updated version of the earlier one. He dedicated it to the memory of his previous book, adding in the introduction that he had *"promised my publishers to write an entirely new book on the same subject on the condition that they subject to auto-da-fé all the remaining copies of the old book."*

Geo moved easily back and forth from writing to research during the thirty years from Mr. Tompkins's first appearance until his own death. As he says in his autobiography:

My major interest is to attack and to solve the problems of nature, be they physical, astronomical or biological. But to "get going" in scientific research one needs an inspiration, an idea. And good and exciting ideas do not occur every day. When I do not have any ideas to work on, I write a book; when some fruitful idea for scientific pursuit comes, writing lags.

Nobody before or since has made the transition back and forth from science writing to research more charmingly or more amusingly.

45   Geo's and Max's Final Messages

Can we now, fifty years later, and armed with new insights on the origin and evolution of life, on the structure and evolution of our cognitive abilities, take a new look at this question and perhaps formulate it in a somewhat less defeatist style? That would seem to me a highly worthwhile undertaking.

Max, commenting on Bohr's and Einstein's approaches

Dear Fremann: This is the crazyest thing I have ever encountered in my life. But where to from here?! Any ideas? —As ever, George

Geo, writing to Freeman Dyson two days before his death

The two comments above are characteristic of Max and Geo, one earnest and intense, the other short and with questioning humor. Their ways of thinking and of acting held true till the end. Max, taking a step back, surveys the field in all its breadth and then decides to move, doing so gently but resolutely. When he said something was a *"highly worthwhile undertaking,"* others listened. Geo, on the other hand, plunged ahead with one more crazy idea, excited about it, as he had been about his many other ones. His strength was waning, but his enthusiasm was not.

Years of excess alcohol consumption meant Geo's health was none too good by the time of the 1965 Penzias-Wilson announcement. Geo described the situation to his friend Alex Rich: *"My liver is presenting the bill for all the drinking."* In recounting this remark to me, Rich said that the alcohol had not seemed to interfere with Geo's reasoning but clearly was affecting his health. As he put it, *"Geo had learned to function despite his heavy drinking; his liver had not."*

Geo remarried in 1958. The union to Barbara Perkins, the publicity manager at Cambridge University Press, proved a happy one. With her help, he managed in his final years to stop his consumption of alcohol, an action further dictated by attacks of hepatitis and other ailments, but it wasn't soon enough to reverse the damage that had been done. He also underwent operations to remove calcium deposits in major arteries. But this did not stop him from working on the physics problems that continued to fascinate him.

Still looking for an explanation for the electromagnetic-to-gravitational-force ratio, Geo wrote to Dyson in October 1967 about his latest thoughts. He described his point of view succinctly: *". . . if one encounters pure big numbers, one has either to accept them or explain them."* It was clear that Geo's preference was and always would be for the latter. Nor did his failing health prevent him from trying to explain those big numbers. In May 1968 he submitted a paper entitled "Observational Properties of the Homogeneous and Isotropic Expanding Universe" to *Physical Review Letters*. The journal published it in June; its last sentence reads, *"More details will be given in a forthcoming paper."* Regrettably, Geo did not live to write that paper.

On August 17, 1968, Geo sent Dyson a renewed attempt he had made to understand the reason for the coincidence of those *"pure big numbers."* It was a draft of a manuscript written with Ralph Alpher entitled "A Possible Relation Between Cosmological Quantities and the Characteristics of Elementary Particles." He sent a note along with it, marked as usual by his terrible spelling, even of Dyson's first name:

Dear Fremann:

   This is the crazyest thing I have ever encountered in my life. But where to from here?! Any ideas?

   As ever, George

Two days later Geo died, still thinking about the expanding universe. He was buried in Boulder's Green Mountain Cemetery. His *"crazyest"* idea, as so many in physics, did not prove fruitful, but I am quite sure that if he were brought back to life, he would be undeterred by its failure and would have even *"crazyer"* ones. Understanding the universe's origins is built out of seemingly crazy ideas.

The final paragraph of the last paper Max wrote is equally revealing. He was revisiting the origins of a famous proposal combining statistical mechanics and quantum theory introduced in 1924 by S. N. Bose and soon developed by Einstein. Commenting on the changes in our view of science, he turned to Einstein's famous saying *"The most incomprehensible thing about the world is its comprehensibility,"* asking if we needed to revisit it in light of later developments. Thinking back on the two great masters of his youth, Bohr and Einstein, Max suggested that their methods for approaching science problems arose from having grown up in a world where the human mind and nature were assumed to be separate entities. According to Max, both Bohr's and Einstein's efforts were like those of *"agonists locked in a peculiar struggle, one trying to ferret out the secrets of Mother Nature and Mother Nature jealously trying to guard them."*

But Max felt this perception had changed. Much had been learned, and he was asking us to rethink our attitude toward nature, not to remain moored to the past:

   Can we now, fifty years later, and armed with new insights on the origin and evolution of life, on the structure and evolution of our cognitive abilities, take a new look at this question and perhaps

formulate it in a somewhat less defeatist style? That would seem to
me a highly worthwhile undertaking.

In that final sentence, Max left us with a message for the future, a hope of making the world more comprehensible by shifting our point of view. It was and still remains a great challenge.

A Festschrift book had been planned for Geo's sixty-fifth birthday in 1969, but his death changed it into a memorial volume. Max wrote the sixteenth of the nineteen contributions. Max's chapter, which is entitled *Fleeting Apparition, Göttingen 1928,* begins with Max's first sighting of Geo:

In the Café Kron & Lanz, in the heart of town, you could sit by the
window on the second floor and watch life go by. Somebody pointed
out to me a slightly sensational figure: a Russian student of theoreti-
cal physics fresh from Leningrad. That was something new.

Almost fifty years to the day after that *"fleeting apparition"* and a decade after Geo died, Max was scheduled to have cardiac bypass surgery to deal with recurring angina pains he had been experiencing. Two days before the surgery, a routine X ray showed that he was suffering from multiple myeloma, a treatable but incurable cancer of the white blood cells. Though it often causes bone erosion and accompanying pain, radiation and chemotherapy made Max comfortable for the remaining three years of his life. He died on March 10, 1981, with Manny at his side.

In thinking about the meaning of an individual's life, Max would often quote a line from Beckett: *"We give birth astride a grave, the light gleams an instant, then there is night once more."* Night had descended on Max, but the memory of him and of Geo lives on.

At this book's beginning, I refer to Max and Geo, somewhat whimsically, as only "ordinary geniuses." But their vision of science's future and what each of them achieved were far from ordinary. The two fields

I recommended to the young student I mentioned earlier, genomics and cosmology, would not have developed the way they have without Max and Geo. What I find most noteworthy about the two, however, was their ability to sense what would become the big areas in science. It is hard enough to identify the most challenging problems of the future when they are in sight, but to have the audacity to want to solve them even before they have been formulated is truly extraordinary. And yet Max and Geo did this and did it more than once.

Max's vision of the gene as a molecular structure obeying the rules of quantum mechanics and his identifying an organism simple enough to trace the way it replicated provided the framework for attacking what came to be known as molecular biology. In the battle he was the strategist, the critic, and the commanding officer, leading a troop that grew over two decades from a few to hundreds. When the battle was finally won, the victory did not come in the guise he had expected, but it was his foresight and his discipline that guided the effort. He was called by some the Pope of the Phage Church, by others the Zen master, and by still others a surgeon of the mind, but there was no doubt that his keen intellect, his spirit, and his joie de vivre fashioned the quest to understand the basic mechanisms that lay at the roots of genetics.

Geo was extraordinary in a different way. His imagination carried him where others could not or would not tread. It took sudden leaps into a dimly shaped future, often skipping all the intermediate steps by which ordinary minds arrived at a conclusion. In some ways one sees this best in what was, after all, a failure: his prediction of a genetic code. To read the Watson and Crick paper and then, with essentially no knowledge of biochemistry, to predict a path for protein formation is truly an extraordinary feat.

Geo habitually gathered bits and pieces of knowledge and then strung them together into astonishing hypotheses about nuclei, stars, galaxies, and, ultimately, about what interested him most, our very universe.

A little while ago, I was strolling across the University of Colorado campus with Geo's son, Igor. After obtaining a Ph.D. in molecular biology, Igor had gone to Caltech as a postdoctoral fellow to work with Max and then returned to Boulder to take a faculty position. As we passed the university library, Igor paused and pointed out to me the inscription over the entrance's portal: *"Who knows only his own generation remains always a child."* He then told me that passing by that very same doorway on a similar walk with Max shortly after Geo's death, Max had commented on the inscription, telling him that remaining *"always a child"* was a good thing for a scientist. One had to approach the world that way, with wonder and eagerness, in order to discover its secrets. And no scientist remained more of a child in the way he set about uncovering nature's secrets than Geo. Until the end he tried to understand in his unique way both what the game was and what its rules were.

Max and Geo educated and encouraged others to come along with them on their adventures in a lighthearted way that many found irresistible. Their doing so changed the way we think of some of the greatest science problems of our modern era in what can rightly be called an extraordinary manner. Each of these "ordinary geniuses" had a special and admirable way of asking: what have we learned in science, and what even more exciting gems may be revealed? It is a heritage we must cherish, a lesson we need to hold dear.

Acknowledgments

Many people have been of great assistance to me in writing this book, but I would like to particularly thank Jonathan Delbrück and Igor Gamow, respectively Max's and Geo's sons, for providing me with their valuable perspectives and with permission to reprint a variety of documents relating to their fathers. Two days that my wife, Bettina, and I spent in Boulder, Colorado, in October 2008 with Igor and his wife, Elfriede, were especially helpful and interesting. In addition, Igor is engaged in a number of creative activities that combine Geo's legacy with his own contributions. I recommend these to the reader.

A number of scientists who knew both Max and Geo shared their memories with me. Among them I would like to single out Freeman Dyson, Alex Rich, Vera Rubin, and James Watson, all of whom graciously met with me and provided me with their insights into the workings of these two fascinating men. An evening with Jim Watson also afforded an inside view of the atmosphere at the Cold Spring Harbor lab that Max helped create and that Watson has done so much to extend. Maureen Berejka was of great assistance in organizing the visit. Other scientists who were helpful are Walter Shropshire, who edited the memorial volume produced on the occasion of Max's centenary; the neuroscientists Lily and Yuh Nung Jan, who were once students of Max's; and finally, Michael Turner, who knows as much as anybody about the ins and outs of cosmology, both the most modern concepts to which he has so greatly contributed and its history in the 1940s and '50s. I also cannot forget Ernst Peter Fischer, who graciously gave me permission to quote extensively from *Thinking About Science: Max Delbrück and the Origins of Molecular Biology*, the wonderfully informative book he wrote with Carol Lipson. Fischer, yet another convert from physics to biology, was, in addition to possessing great writing skills, a graduate student of Max's in the 1970s.

A number of historians of science were also very helpful. I would like to single out Phillip Sloan, who has studied in great detail Max's contributions to biology in the mid-1930s and who will be presenting them in a forthcoming book from the University of Chicago Press. Finn Aaserud welcomed me in Copenhagen at Bohr's Institute and showed me once again what the Copenhagen spirit was and still is. I also enjoyed meeting Helge Kragh in Aarhus, and studying his learned *Cosmology and Controversy*. Karl Hufbauer was very gracious in providing information about Geo and also about Russian physics as it related to cosmology in the tumultuous 1930s.

I would also like to thank Shelley Erwin and Loma Karklins at the California Institute of Technology Archives, Jennifer Kinniff at George Washington University's Gelman Library, the assistants at the Library of Congress, Joseph Anderson and Gregory Good at the Center for History of Physics of the American Institute of Physics, and, once again, Felicity Pors at Bohr's Institute in Copenhagen.

My wife, Bettina, patiently offered her own views on the many drafts of this book, greatly improved both the style and the organization of its contents, put up with my grumblings, and unfailingly offered love and support during the course of both the writing and the research. Three friends, John Breit, Paul Broda, and Peter Sterling, read the manuscript in its entirety and offered valuable criticism, though of course any remaining errors are solely my own responsibility.

I am grateful to my agents Katinka Matson and John Brockman for their creative approach to making the world of science better known to the wider public and for helping me to contribute to this cause. My wonderful editor at Viking, Wendy Wolf, once again encouraged me in the book's initial stages and then placed me in the capable hands of Alessandra Lusardi, who through successive detailed readings helped reshape the manuscript into what I hope is a readily accessible form. Her work has reminded me how valuable a truly good editor can be.

Finally, I would like to thank Brittney Ross at Viking, who helped greatly in the final stages.

Notes

Abbreviations

BSC   Niels Bohr Scientific Correspondence. Niels Bohr Institute, Copenhagen, Denmark.

DCTech   The Max Delbrück Archives, California Institute of Technology.

Geo-MWL   Gamow, George. 1970. *My World Line: An Informal Autobiography*. New York: Viking.

GWUEncy   *The George Washington University and Foggy Bottom Historical Encyclopedia*. Online resource. At http://encyclopedia.gwu.edu/gwencyclopedia/index.php?title=Main_Page.

OH-Geo   Oral history transcript. American Institute of Physics. Interview with George Gamow, by Charles Weiner at Professor Gamow's home in Boulder, Colorado, on Apr. 25, 1968.

OH-Max   Oral history transcript. Interview with Max Delbrück by Carolyn Harding, July 14–Sept. 11, 1978. Archives of the California Institute of Technology.

Patoomb   Cairns, John, with Gunther Stent and James Watson, eds. 1966. *Phage and the Origins of Molecular Biology*. Plainview, N.Y.: Cold Spring Harbor Laboratory Press.

WMAP   Wilkinson Microwave Anisotropy Probe. National Aeronautics and Space Administration.

Introduction

xviii   *"kept his eye on . . . can do"*: Hershey (1981), as quoted by Fischer and Lipson (1988), p. 150.

xviii   *"Max's attitude was Dionysian . . . a moveable feast"*: Luria (1984), p. 129.

xix   *"I still cannot . . . than anyone else"*: Inglis (2003), p. 143.

xix   *"so very often . . . everybody"*: J. D. Watson, private discussion with author, July 12, 2009.

xx   *"I like the pioneering thing"*: OH-Geo, p. 75.

xx   *"I decided to choose myself . . . to cosmology"*: Ibid.

xxi *"Ah, the two mavericks!"*: Rose Bethe, private discussion with author, Sept. 10, 2008.

1. When Max and Geo First Met

3 *"I found out at an early age . . . you felt at home"*: Delbrück (1972), p. 133.

2. Max Grows Up

6 *"Germans had their backs . . . military service"*: Bethge (1970), p. 17.
6–7 *" 'Do you know that your' . . . under control"*: Von Meding (1997), p. 4.
7 *"If you are ever uncertain . . . right thing"*: Ibid., p. 6.
7 *"one will, one force . . . petty and common"*: Heilbron (1986), p. 80.
7 *"relatively affluent . . . a ghost town"*: OH-Max, p. 9.
10 *"The underlying . . . complex to be solved"*: Dirac (1929), p. 714.
11 its leading exponent, at twenty-six, was still a very young man: Eugene Wigner was awarded the 1963 Nobel Prize in Physics for the work he began at Göttingen in the 1920s and later extended.

3. Geo Grows Up

16 *"In more than twenty years . . . a practical joke"*: Watson (2001), p. 90.
17 Each of the three later had an extraordinary career: A fourth physics student, Matvei Bronstein, frequently joined the Three Musketeers. The quartet was known as the Jazz Band. Bronstein was arrested in the Great Purge of 1937, tried, and executed.

4. Göttingen and Copenhagen

20 *"as if you fired . . . and hit you"*: Andrade (1964), p. 111.
20 *"The explanation did not appeal . . . in this case"*: Geo-MWL, p. 60.
21 *"a somewhat mystical . . . Copenhagen"*: French and Kennedy (1985), p. 60.
22 *"The task of having to introduce . . . of science"*: Pais (1991), p. 171.

5. Particle or Wave?

27 *"Anyone present . . . through the window"*: Geo-MWL, p. 60.
28 *"My secretary . . . for one year"*: Ibid., p. 64.
29 as did a Göttingen friend of his, Fritz Houtermans: Geo's friend Fritz Houtermans was a chain smoker who preferred working in cafés, where he would consume one cup of coffee after another. An increasingly ardent Communist, he left Germany for England when Hitler came

to power and from there immigrated to the Soviet Union shortly after Geo, increasingly disillusioned with Stalinist Russia, fled to the United States. Geo's decision wound up being the far wiser one; Houtermans was imprisoned by the Soviet secret police in 1937. He was tortured and accused of being both a Trotskyite and a German spy. Released to the Germans in 1940 after the signing of the Hitler-Stalin Pact, he was promptly imprisoned by the Nazis as a Soviet spy, but eventually freed. He died in 1966.

30 *"Build me . . . without any trouble"*: Geo-MWL, p. 83.
30 *"The discovery of the radioactive . . . of evolution"*: Eve (1939), p. 120.

6. Max's and Geo's Early Careers

33 *"I wrote one paper . . . I wasn't ready"*: OH-Max, p. 36.
34 *"I had not felt . . . carry on with"*: Ibid., p. 34.
36 *"In accordance . . . and electrons"*: Casimir (1983), p. 117.
37 *"It has never . . . will undoubtedly do"*: Ibid. p. 117

7. Copenhagen, 1931

38–39 *"We went to the flicks . . . really proved"*: Casimir (1983), p. 98.
39 *"I am grateful to drs. R. E. Peierls . . . July 25"*: Gamow (1930), p. 131.
40–41 *"The note by Guido Beck . . . misunderstandings"*: Reines (1972), p. 283.
41 *"Of course the paper . . . discontinue them"*: Ibid., p. 283.
42 *"I really longed . . . vital force"*: OH-Max, p. 47.
43 *"I cannot go to Blegdamsvej"*: Gamow (1966), p. 201.

8. Zurich, 1931

45 *"in any case better . . . quantum mechanics"*: Enz (2002), p. 250.
45 *"Yes, I know you are fond . . . mathematics"*: Van der Waerden (1967), p. 3.
46 *"I still feel a strong friendly . . . characters"*: Enz (2002), p. 352.
46 *"Look, Max . . . boring papers"*: OH-Geo, p. 92.
46 *"It also motivated . . . dreamed of before"*: OH-Max, p. 40.
47 *"he would be forced . . . for him"*: Pauli (1985), vol. 2, letter 291, p. 114.
47–48 *"I have accepted . . . für Biologie"*: As quoted in Aaserud (1990), p. 95.

9. Max, Bohr, and Biology

49 *"that one would . . . own father"*: Shropshire (2007), p. 79.
49–50 *"It was his task . . . inorganic nature"*: As quoted by Aaserud (1990), p. 70.
51 *"the special . . . whole of science"*: Bohr (1972), vol. 6, p. 101.

51 "Before I conclude . . . here described": Ibid., p. 251.
52 "He [Bohr] talked . . . and biology": OH-Max, p. 41.
53 "completely convinced . . . is defending": Moore (1989), p. 228.
53 "very careful never . . . follow it up": OH-Max, p. 43.
53 "we should doubtless . . . in biology": Bohr (1933), p. 458.
54 "infinitely more competent": OH-Max, p. 38.
55 "very learned paper": OH-Max, p. 44.
55 "This semester I have . . . learn a lot": Bohr (1972), vol. 10, p. 467.
56 "complained bitterly . . . Jordan's views": Ibid., p. 466.

10. Max, Berlin, and Biology

58 Activities in his laboratory . . . took off rapidly after that: The analysis
    of *Drosophila* mutants has continued to the present. Nor is this mere
    classification, for the importance of studying them has remained
    unabated. The original classics, *"white, yellow, singed, and miniature,"*
    have been joined by others, described by their anomalous appearance
    (curly, freckled, or furry), their curious behavior (breathless, grim), or
    perhaps their slowness in learning (amnesiac, dunce, and rutabaga).
    There is even a mutant with no heart, tinman.
59 "meet the physicist . . . hopelessly vitalistic": As quoted by Aaserud
    (1990), p. 98.
60 "we talked, usually . . . during the session": Patoomb, p. 37.
60 "three dimensional . . . hold for experiment": Jacob (1982), p. 225.
60 "somewhat insolent . . . piece of butter": Ratner (2001), p. 933.
61 "arrive at a theory . . . concepts of physics": Sloan and Fogel (2011), p. 37.
61 "the view of gene mutation . . . considered secure": Ibid., p. 41.
62 "The paper contains . . . developmental process": Bohr (1972), vol. 10,
    p. 472.

11. Geo Escapes from Russia

63 "A son of the working . . . of an atom": Geo-MWL, p. 74.
64 "Gamow has kept away . . . bohemia": As quoted by Kragh (1996), p. 90.
64 "Nothing of interest . . . after that": Igor Gamow, private communication
    with author.
64–65 "I had always felt . . . heredity": Geo-MWL, p. 120.
67 "very childish . . . the whole affair": Ibid., p. 111.

12. The Russia Geo Left Behind

70 "As a result of the . . . this category": Spruch (1979), p. 36.
70 "Don't you know . . . won't sing": Ibid., p. 36.

70   *"This bird will sing"*: Ibid., p. 36.

70–71   *"is a trouble-maker . . . writing to you"*: Khalatkinov (1989), p. 147.

71   *"I am interfering . . . Soviet Union"*: Ibid., p. 147.

71   *"Everyone knew . . . tiger's cage"*: Ibid., p. 147.

72   Geo always regarded "Dau" as his dearest friend: Kapitza, Landau's protector, survived the Stalinist purges unscathed because his skill in experimental work and in building scientific instruments was of great importance for Soviet science and technology. But in 1946, after refusing to work on nuclear weapons, he, too, fell into disfavor. He spent the next decade under house arrest, his condition only easing after Stalin's death. In 1978 he received the Nobel Prize in Physics, an honor bestowed on Landau in 1962.

73   *"Good, that will . . . Timofeev"*: As quoted by Fischer and Lipson (1988), p. 275.

73   *"It brings back . . . riddles of life"*: DCTech, box 22, folder 3.

73   Timofeev-Ressovsky . . . now languishing under the rule of Soviet dictators: While in the gulag, Timofeev shared a cell with Alexander Solzhenitsyn for a while. The great Russian author described in *The Gulag Archipelago* how Timofeev organized science seminars even while in prison and later used him as a model for the scientist in *The First Circle*.

13.   Geo Comes to America

76   *"At the crack of dawn . . . right and new"*: Harper (1997), p. 125.

14.   The Sun's Mysteries Revealed

81   *"Well, I've always been . . . is possible"*: Chandrasekhar (1977), p. 12.

82   *"I was totally uninterested . . . in my life"*: Harper (1997), p. 46.

82–83   *"several specific . . . groups represented"*: GWUEncy (1938).

83   *"Well, it's the kind . . . on this"*: Bethe (1966), no page numbers in transcript.

83   *"for his contributions . . . in stars"*: Nobel Prize citation, 1967.

84   *"most remarkable . . . Gamow's game"*: Blumberg and Panos (1990), p. 94.

85   *"they paid me $100 . . . contribution"*: OH-Geo, p. 83.

15.   Max Leaves Germany

90   *"After the boycott . . . her health"*: Von Meding (1997), p. 9.

91   *"thorough and rigorous . . . Socialist State"*: Hentschel (1996), p. xxix.

92   *"checking up on . . . were doing"*: OH-Max, p. 56.

92   *"Don't you want . . . these people"*: Ibid., p. 56.

94   *"to study the theory . . . physical agencies"*: As quoted by Fischer and Lipson (1988), p. 97.

16. Max in the New World

98   d'Herelle called the viruses bacteriophages: Though d'Herelle was a remarkable research scientist, his primary interest was in developing treatments for ailments such as cholera. He met with some success in developing phages that attacked disease-causing bacteria, but the results were mixed, since the bacteria often developed resistance to the phages. Though d'Herelle tried to counter this with phage cocktails, their production was erratic and the results often inconsistent. Phage therapy, as it was known, was largely abandoned in the 1940s after the development of antibiotics. However, it may be making a comeback in treating bacterial disease because of both its greater specificity and the emergence of antibiotic-resistant bacteria.

99   *"Such motives . . . our cause"*: Anderson (1945), p. 264.

17. Fission

103   *"Bohr has gone crazy . . . nucleus splits"*: Blumberg and Panos (1990), p. 46.

104   *"Our radium . . . like barium"*: Sime (1996), p. 233.

105   *"Oh but this is wonderful . . . about it"*: Frisch (1979), p. 116.

107   *"Let's run this experiment tonight"*: Tuve (1982), no page numbers in transcript.

108   *"The Department of Physics . . . last two years"*: Weart and Szilard (1978), p. 62.

109   contains a small admixture of this rare form: A uranium nucleus always has ninety-two protons, but the number of neutrons it contains can vary. This is a feature of all elements; though every hydrogen atom, for example, has a nucleus with a single proton, one or two neutrons may accompany the proton. Those hydrogen atoms with neutrons—considerably rarer than the ordinary form, which has none—are called deuterium and tritium. They are hydrogen isotopes; the three forms are chemically identical but have different masses. In the same way, though more than 99 percent of commonly found uranium has 136 neutrons, approximately 0.7 percent has only 133. Labeling the two forms by atomic weight, the sum of protons and neutrons, one is called U-238 and the other U-235.

110   *"it is conceivable . . . be constructed"*: Kelly (2007), p. 43.

110   *"It would have been . . . after Hiroshima"*: Geo-MWL, p. 148.

18. Supernovae and Neutron Stars

111 whole stars exploding: The first atom bomb test was conducted on July 16, 1945, at a secret test site near Alamogordo, New Mexico. Those present were in awe of the weapon's power. As the great physicist Isidor Rabi remembered, *"Suddenly there was an enormous flash of light, the brightest light I have ever seen or that I think anyone has ever seen"* (Rhodes [1988], p. 672). The flash's intensity led an early popular book about the atom bomb project to take the title *Brighter Than a Thousand Suns*. But supernovae, the explosions Geo was thinking about in 1939, were brighter than a billion suns.

112 *"More than a half-century . . . fusion chains":* Raffelt (1996), p. 1.

115 *"the Urca Process . . . Casino da Urca":* Geo-MWL, p. 147.

116 *"for pioneering contributions . . . neutrinos":* Nobel Prize citation, 2002.

19. Max Meets Manny and Sal

119 *"that year among . . . way of biologists":* Luria (1984), p. 19.

119 *"Between bacteriophage . . . at first sight":* Ibid., p. 20.

119–20 *"for the first time . . . about research":* Ibid., p. 20.

121 *"Pauli simply asked . . . not a word":* Ibid., p. 33.

122 *"He couldn't wait . . . Cold Spring Harbor":* As quoted by Fischer and Lipson (1988), p. 136.

124 *"the standard by which . . . be measured":* Stent (1981), p. iii, as quoted by Fischer and Lipson (1988), p. 146.

20. Hitting the Jackpot

125 the electron microscope: Ernst Ruska was awarded the 1986 Nobel Prize in Physics for the development of the electron microscope.

126 *"They formed a little . . . his paradise":* Patoomb, p. 63.

128 *"Goliath of physical chemistry":* Luria (1984), p. 74.

21. What Is Life?

132 *"is usually expected . . . of ourselves":* Schrödinger (1944), p. 1.

133 *"there is no alternative . . . further attempts":* Ibid., p. 61.

133 *"Delbrück's molecular . . . and genetics":* Ibid., p. 72.

133–34 *"No detailed information . . . as the former":* Ibid., p. 72.

134 *"My change of heart . . . living world":* Watson (2003), p. 35.

135 *"it was only later . . . around the corner":* Crick (1988), p. 18.

22. The Phage Group Grows

136 *"Twenty-five years ago. . . this promised land"*: Patoomb, p. 178.

136 *"Drinks whiskey . . . likes independence"*: Ibid., p. 175.

137 *"two enemy aliens and a social misfit"*: As quoted by Fischer and Lipson (1988), p. 149.

137 *"The Phage Church . . . was the saint"*: Stahl (2001), p. 144.

138 *"One spring evening . . . immediately"*: Patoomb, p. 134.

138 *"In that three-week course . . . to listen"*: Ibid., p. 134.

139 *"a tremendous vital force"*: OH-Max, p. 37.

139 *"reluctance to enter . . . evidently enjoyed"*: Luria (1984), p. 129.

140 *"As the summer passed . . . got sorted out"*: Patoomb, p. 241.

141 *"Delbrück wanted to . . . all in one"*: Perutz (1998), p. 180.

141 *"Unlike Delbrück . . . field of science"*: Brenner (2007), p. 1246.

142 *"the parents . . . exchanged something"*: Delbrück (1948), as quoted by Fischer and Lipson (1988), p. 168.

143 *"I refuse to believe . . . clean this up"*: Feb. 26, 1948, letter from Max Delbrück to Joshua Lederberg. The Joshua Lederberg Papers, National Library of Medicine, Bethesda, Maryland.

143 *"for his discoveries . . . in bacteria"*: Nobel Prize citation, 1958.

23. Geo and the Universe

145–46 *"It would be so nice . . . (heredity of course!)"*: BSC, Oct. 24, 1945, letter.

146 Geo knew . . . what such a blast's effects might be: I recently found an old letter to my father from my uncle Emilio, who had worked on the atom bomb, advising my father on what constituted a safe distance away from a metropolitan center that might be a target. He suggested my father might want to move his family from Manhattan.

146 *"Einstein would meet me . . . comments"*: Geo-MWL, p. 149.

147 *"I was sitting . . . theory of gravitation"*: Pais (1982), p. 179.

147 *"the happiest thought of my life"*: Ibid., p. 178.

148 *"Matter tells space . . . how to move"*: Misner (1973), p. 5.

150–51 *"The subject that fascinated . . . his lectures"*: Geo-MWL, p. 41.

24. Gamow's Game

153 *"I should be willing to pay . . . an hour"*: Johnson (2005), p. 32.

154–55 Sir Arthur Eddington . . . static model of the universe: Eddington neglected, however, to mention Lemaître's obscure 1927 paper in his 1930 presentation. It had probably been forgotten or unread by him. After Lemaître wrote him a note enclosing a copy of it, Eddington moved quickly to credit Lemaître, translated the note he had been forwarded, and saw to its publication in the leading British astronomy journal.

155 *"the biggest blunder he ever made in his life"*: Geo-MWL, p. 44.

155 *"I don't think that . . . is static"*: Weinberg (2005), p. 31.

156 *"The idea that . . . given to us all"*: As quoted by Kragh (1996), p. 59.

156–57 *"stellar thermonuclear reactions were Gamow's game"*: Blumberg and Panos (1990), p. 94.

157 *"no elements heavier . . . ordinary stars"*: Bethe (1939), p. 434.

157 where did they come from: Wanting to have a discussion on how elements might be formed in the early universe, Geo chose "The Problems of Stellar Evolution and Cosmology" as the topic for the Eighth Annual Conference on Theoretical Physics, held in Washington, D.C., Apr. 25–28, 1942. At the meeting, Chandrasekhar described his calculations (with Louis Henrich) of the element abundance expected from a hot and dense early universe. The results were not encouraging. The best fit to the available data was obtained by assuming a primeval gas in thermal equilibrium at a temperature of almost ten billion degrees centigrade and a density of ten million grams per cubic centimeter, ten million times denser than water. There was some agreement of theoretical predictions with experimental data pertaining to elements up to and including argon, number 18 in the 92-member periodic table, but complete failure after that. Some key ideas were clearly missing.

157 *"coagulating into . . . atomic species"*: Gamow (1946), p. 573.

157 *"at the epoch . . . one second"*: Ibid., p. 573.

25. Bohr, Geo, and Max

161 *"discovery of the . . . by X-ray irradiation"*: Nobel Prize citation, 1946.

161–62 *"His [Muller's] work . . . the next"*: Watson (2003), p. 41.

163 *"hostility towards chemistry . . . to do so"*: OH-Max, p. 73.

164 *"ideas were taken . . . are essential"*: Dyson (1985), p. 6.

26. Back to Germany

167 *"I want to add . . . like him so much"*: Pauli (1994), letter 615.

168 *"nothing except . . . with my father-in-law"*: Von Meding (1997), p. 23.

168 *"I lived with . . . preparing a meal"*: Ibid., p. 28.

169 *"If anybody feels guilty . . . Otto Hahn certainly"*: OH-Max, p. 103.

169 *"You all worked . . . to help you"*: Sime (1996), p. 310.

170 *"And I hope, dear . . . and Germany"*: Schweber (2000), p. 95.

170 *"may have influenced . . . and the world"*: Shropshire (2007), p. 153.

27. The New Manchester

173 *"the inflexible . . . minimize it"*: Kay (1993), p. 209.

173 Max and Pauling were also friends: Max and Pauling not only knew each other but had even written a paper together in 1940, a response to another of German physicist Pascual Jordan's ventures into biology. Jordan's new foray was based on a supposed quantum mechanical attraction between identical molecules. Max didn't believe the conjecture but, to be safe, he consulted Pauling. The two were in total agreement and wrote a short article for *Science* criticizing Jordan's work. But the most interesting feature of the note is a proposal of their own that chemical stability might instead be achieved by *"a system of two molecules with complementary structures"* (p. 79).

173 *"because I believe that . . . in the 1910s"*: BSC, Jan. 11, 1947.

174 *"signals the completion . . . into a biologist"*: Ibid.

176 *"At night there was . . . Max at the helm"*: Shropshire (2007), p. 153.

28. Alpha, Beta, Gamma

178 *"from Stettin . . . the Continent"*: Churchill (March 5, 1946), http://en.wikisource.org/wiki/Iron_Curtain_Speech.

179 *"I felt at the time . . . being added"*: Alpher (1988), p. 28.

182–83 *"You'd have to be . . . 'It isn't even wrong'"*: Turner (2008), p. 8.

29. Big Bang Versus Steady State

187 *"to fellow cosmogonists . . . and ages"*: Gamow (1952), cover page.

187 *"primordial Fiat Lux . . . of galaxies"*: Translated by McLaughlin, as quoted by Kragh (1996), p. 257.

187 *"It can be considered . . . truth that"*: Gamow (1952a), p. 251.

188 *"In view of . . . 'Parisian fashion'"*: Gamow (1952), p. 1.

188 *"to assume that . . . pressure conditions"*: Ibid., p. 56.

188 *"inexperienced housewife . . . for the pie"*: Ibid., p. 56.

189 *"Except, of course . . . up to speak"*: Hoyle (1994), p. 256.

190 *"humanity . . . different topics"*: Salpeter (1996), p. 14.

191 *"delivered brilliant . . . I have known"*: Rubin (2002), p. 13.

30. DNA

192 *"the model . . . my own life"*: Watson (2000), p. 213.

192 *"The fox knows . . . big truth"*: Attributed to the Greek poet Archilocus. "The Hedgehog and the Fox" also served as the title for a celebrated essay by the British philosopher Isaiah Berlin.

196 *"If we are . . . implications"*: Patoomb, p. 186.

196 *"It is a lot of fun . . . else tries to"*: Ibid., p. 187.
197 *"If the results . . . undetermined"*: McCarty (1985), p. 155.

31. The Double Helix

200 *"The secret of life . . . to solve it"*: Perutz (1998), p. 181.
201 *"One day in September . . . find it"*: Ibid., p. 188.
201 *"Often he came up . . . on theory"*: Watson (1968), p. 10.
202 *"their discoveries . . . living material"*: Nobel Prize citation, 1962.
203 *"it has not escaped . . . genetic material"*: Watson and Crick (1953), p. 738.
204 *"We would prefer . . . him a copy"*: Watson (1968), p. 136.
204 *"Delbrück hated . . . secrecy in scientific matters"*: Ibid., p. 127.
204 *"Very remarkable . . . Rutherford in 1911"*: Bohr (1972), vol. 10, p. 474.
204 *"I have a feeling . . . tumultuous phase"*: Delbrück letter to J. D. Watson, Apr. 14, 1953; original is in Delbrück papers, California Institute of Technology Archives. It is also quoted by Fischer and Lipson (1988), p. 201.
204–5 *"The great thing . . . grand romantic manner"*: Medawar (1968), p. 3.
205 *"I don't think Delbrück cared . . . there might be"*: Crick (1988), p. 61.
205–6 *"'Nobody, absolutely . . . for everyone'"*: Judson (1996), p. 41.
206 *"always thinking . . . and politics"*: James Watson, private discussion with author, July 12, 2009.
206 *"Can you patent it?"*: Watson (2003), p. 58.
207 *"Since the two chains . . . be insuperable"*: Watson and Crick (1953a), p. 966.
207 *"Increasingly, by then . . . gracefully reversed course"*: Watson (2001), p. 24. Though Max was graceful in reversing course, he was always reluctant to accept a statement as true until it had been decisively shown to be correct. Some were hurt by his obstinacy and some cowed, but most of those who knew him well appreciated his directness. If nothing else, it provided a testing ground for their convictions. Max had seen Bohr act this way. Abraham Pais, Bohr's biographer and a well-known physicist himself, remembered a conversation with Bohr where *"Bohr and I talked about an event when a senior theoretical physicist had talked a younger colleague out of publishing a result that turned out to be correct and important. When I remarked that this was a sad story, Bohr literally rose and said: 'No, the young man was a fool.' He explained that one should simply never be talked out of anything one is convinced of"* (Pais [1991], p. 239). Luckily for them, neither Watson and Crick nor Meselson and Stahl were fools. They had not allowed Max to talk them out of their discoveries.

32. Geo and DNA

209 *"Look what a wonderful . . . have written"*: OH-Geo, p. 93.
210 *"I am very much . . . exact sciences"*: Watson (2001), p. 24.

213 *"The idea that . . . ideas explicitly":* Crick (1988), p. 92.

213–14 *"The point, the contribution . . . might exist":* Judson (1996), p. 281.

214 *"because, of course . . . unhappy about it":* OH-Geo, p. 93.

214 *"Best regards from Geo":* Watson (2001), p. 60.

215 *"I am still nuclear physicist . . . H bombs":* BSC, Feb. 1, 1954.

216 *"Do or die, or don't try":* Watson (2001). See George Gamow Memorabilia, p. 3.

216 a friend who lived in Woods Hole: Geo's friend who lived in Woods Hole was Albert Szent-Györgi, whom he had known in Cambridge, England, during the late 1920s. Szent-Györgi won the 1937 Nobel Prize in Physiology or Medicine for his work on vitamin C, and was now studying the biophysics and biochemistry of muscle action. As his name suggests, he was a Hungarian (appropriately enough paprika was used in his studies of vitamin C) but he left Hungary in 1947, unhappy about the Russian takeover, and settled in Woods Hole.

217 *"On most afternoons . . . it is now":* Crick (1988), p. 94.

217 *"Particularly lethal . . . whisky":* Watson (2001), p. 91.

217 RNA Tie Club: Judson (1996), p. 269.

33. Geo Begins Again

220 *"We must take . . . its arbitrariness":* Fermi (1962), p. 720.

220 *"The solution looks . . . inelegant":* Geo-MWL, p. 148.

220 *"Whenever a spontaneous . . . more efficient":* Unpublished remark, commonly attributed to Leslie Orgel.

220 *"Evolution is cleverer than you are":* Ibid.

220 *"The genetic code . . . product of accident":* Crick (1988), p. 101.

221 *"it is obvious . . . about it":* Judson (1996), p. 289.

222 *"What do you think . . . amino acids":* Watson (2001), p. 187.

222 *"Here of course . . . structure is wrong":* Olby (1974), p. 431.

34. Max Begins Again

226 *"We all spoke German . . . tower":* Shropshire (2007), p. 164.

227 *"Guten Tag . . . is horrible":* Ibid., p. 91.

227 *"would seem wonderfully fitting . . . this direction":* Bohr (1972), vol. 10, p. 488.

227 *"that the very . . . atomic physics":* French and Kennedy (1985), p. 319.

35. The Molecular Biology That Was

229 *"annual phage meetings . . . of participants":* Patoomb, p. 178.

229 *"more and more exciting . . . to my wife":* Ibid., p. 165.

229 *"Dear Dotty: please . . . is important":* Ibid., p. 153.

230 *"Max lost interest . . . molecular"*: Alex Rich, private communication with author, June 28, 2008.
231 *"In biology . . . ill-used term of biophysics"*: Patoomb, p. 23.
231 *"like a child's . . . dime store"*: Judson (1996), p. 41.
232 *"Max was a surgeon . . . own hands"*: Shropshire (2007), p. 145.

36. The Phage Church Trinity Goes to Stockholm

235 *"While the scientists. . . . 'The Revenge of Truth'"*: Delbrück Nobel Prize speech, 1969.
235 *"any lie . . . become true"*: Dinesen (1934), p. 122.
235 *"'The truth, my children . . . consequences'"*: Ibid.
236 *"A master who lived . . . replied the master"*: Zen Buddhism (1959), parable number 1.
236 *"One is what . . . at least"*: Beckett (1955), p. 72.
236 *"in order to blacken . . . without incident"*: Beckett (1955), p. 92.
236–37 *"playing animals . . . play together"*: Delbrück (1972), p. 151.

37. The Triumph of the Big Bang

242 *"A possible explanation . . . this issue"*: Penzias and Wilson (1965), p. 420.
242 *"The presence of thermal . . . ten billion degrees"*: Dicke et al. (1965), p. 416.
242 *"this was the type . . . Herman and others"*: Dicke et al. (1965), p. 418.
242 *"It is very nicely . . . almighty Dicke"*: Reines (1972), p. 35.

38. The Cosmic Microwave Background Radiation

245 average mass density . . . different from what we now believe: The Big Bang should be viewed as an explosive expansion of space itself rather than as an explosion occurring at a particular site. The distance between any two points is therefore growing. Since the wavelength of radiation constitutes a measure of length, we must take it as increasing. It does so by a factor of a thousand between the time of atom formation and the present; hence a photon's energy, known to be inversely proportional to its wavelength, decreases by that very same factor.

The cosmic microwave background radiation is therefore a snapshot of photon distribution at a time when the ambient temperature was three thousand degrees, but with each photon's wavelength decreased by a factor of a thousand. Hence it appears as a distribution at three degrees. This does *not* mean the present universe is in thermal equilibrium at three degrees. We are only seeing a record of the last time it was in such a state, some four hundred thousand years after the Big Bang.

246 *"I think most importantly . . . nuclear physics"*: Weinberg (1977), p. 131.

39. Cosmology's New Age

248 *"for their discovery . . . background radiation"*: Nobel Prize citation, 1978.

251 *"for their discovery . . . background radiation"*: Nobel Prize citation, 2006.

252 *"All the arguments . . . one fell swoop"*: WMAP Web page, http://map .gsfc.nasa.gov.

252 *"WMAP has started . . . first time"*: Ibid.

40. Einstein's Biggest Blunder

254 *"the biggest . . . his life"*: Geo-MWL, p. 44.

255 *"All things are numbers"*: Kahn (2001), p. 27. For an easily accessible version of Pythagorean thought, see Koestler's book. For a more scholarly presentation, see Kahn's.

256 *"There's not the . . . cannot hear it"*: Shakespeare (1992), p. 2.

41. Duckling or Swan?

259 *"The garbage . . . of the present"*: Zee (2003), p. 498.

261 *"That would be an exaggeration . . . to be close"*: OH-Max, p. 36.

261 *"shy as a gazelle . . . Victorian maid"*: *Sunday Dispatch*, Nov. 19, 1933, as quoted by Segrè (2007), p. 83.

262 *"The first criticism of this idea . . . 'get married'"*: Gamow (1967a), p. 766.

262 *"Well, the increased supply . . . got an idea"*: Ibid., p. 766.

262–63 *"Well, the old horse . . . told myself"*: Ibid.

42. After the Golden Age

264 *"DNA makes . . . protein"*: Crick (1988), p. 109.

266 Wally Gilbert found one: My wife, Bettina, and I were recently having breakfast in a Cambridge, Massachusetts, restaurant when Bettina commented, as she sometimes does, that a person at a nearby table had an interesting face. I turned around and recognized Wally Gilbert. Bettina and I went over and started chatting with him and his daughter. I was enormously pleased to hear that he had read and liked *Faust in Copenhagen*, the book I had recently written. But given who Wally is, I suspected there might be more to come and was prepared for some surprises.

He had, as I described earlier, started his career as a physicist and then switched to molecular biology, where he made a major discovery by isolating the repressor gene. After receiving the Nobel Prize in Chemistry for his pioneering work in gene sequencing, Wally decided it might be interesting to go into business. He became a co-founder and the chairman of Biogen, one of the first biotech start-up companies. After some time in the finance world, he returned to Harvard, once again a key figure in research, including studies of how life might have originated. But that was then. Wally now told us that he had decided a few years ago to give up science and concentrate on artwork, using digital photography as a medium. I was reassured to see that the *"pioneering thing"* is alive and well, at least in certain hands.

By the way, I do not regret having stayed in theoretical physics.

267 *"for their contributions . . . nucleic acids"*: Nobel Prize citation, 1980.
267 *"Had anyone suggested . . . another drink"*: Watson (2003), p. 193.

43. The Unavoidable and the Unfashionable

270 *"Now that the success . . . research"*: Patoomb, p. 8.
271 *"You, your joys . . . molecules"*: Crick (1994), p. 3.
271 *"come to . . . of consciousness"*: Ibid., p. 13.
271 *"It's not so much that he is good . . . same way"*: Judson (1996), p. 45.
271 *"what do we know . . . do we know it?"*: Delbrück (1986), p. 1.
272 *"That a man should . . . question"*: Ibid., p. 21.
272–73 *"being a finite engine . . . be understood"*: Patoomb, p. 8.
274 *"Both strived to do . . . the crowd"*: Yuh Nung Jan, private communication with author, Sept. 4, 2009.
274 *"when Delbrück . . . molecular biology"*: Patoomb, p. 178.
275 *"Space, cosmology . . .Victorian"*: Freeman Dyson, private communication with author, Aug. 12, 2009.
275–76 *"In recent . . . eyes open"*: by permission of Dr. Freeman Dyson.
276 *"'I am charting . . . I think?'"*: Shropshire (2007), p. 155.

44. Mr. Tompkins Arrives

278 *"'The time has come . . . of many things'"*: Carroll (2009), p. 1.
278 *"Above all my thanks . . . intended it to be"*: Gamow (1947), p. 1.
279 *"a little clerk . . . bank"*: Gamow (1939), p. 1.
279 *"Your article will be published . . . more. Snow"*: Geo-MWL, p. 156.
280 *"promised my publishers . . . old book"*: Gamow (1964), p. 1.
281 *"My major interest . . . writing lags"*: Geo-MWL, p. 160.

45. Geo's and Max's Final Messages

283 *"My liver is . . . drinking"*: Alex Rich, private communication with author, June 28, 2008.
283 *"Geo had learned to function . . . his liver had not"*: Ibid.
283 *"if one encounters . . . explain them"*: Freeman Dyson, private communication with author, Aug. 12, 2009.

283   *"More details . . . paper"*: Gamow (1968), p. 1312.
284   *"This is the crazyest . . . Any ideas?"*: Ibid. Freeman Dyson, private communication with author, Aug. 12, 2009.
284   *"The most incomprehensible . . . comprehensibility"*: Vallentin (1954), p. 24.
284   *"agonists locked . . . guard them"*: Delbrück (1980), p. 470.
284–85   *"Can we now . . . undertaking"*: Ibid.
285   *"In the Café Kron & Lanz . . . something new"*: Reines (1972), p. 280.
285   *"We give birth astride . . . once more"*: As quoted by Fischer and Lipson (1988), p. 295.
287   *"always a child"*: Igor Gamow, private communication.

Bibliography

Aaserud, Finn. 1990. *Redirecting Science: Niels Bohr, Philanthropy and the Rise of Nuclear Physics.* Cambridge, UK: Cambridge University Press.

Alpher, R. A., and R. Herman. 1948. "Evolution of the Universe." *Nature* 162 (1948): 774–75.

Alpher, Ralph A., with Hans Bethe and George Gamow. 1948. "The Origin of Chemical Elements." *Physical Review* 73 (1948): 803–4.

Alpher, Ralph A., with James Follin and Robert Herman. 1953. "Physical Conditions in the Initial Stages of the Expanding Universe." *Physical Review* 92 (1953): 1347–61.

Alpher, Ralph A., with George Gamow and Robert Herman. 1948. "Thermonuclear Reactions in an Expanding Universe." *Physical Review* 74 (1948): 1198–99.

Alpher, Ralph A., with Robert Herman. 1988. "Reflections on Early Work on Big Bang Cosmology." *Physics Today* (Aug. 1988): 24–34.

Anderson, Thomas, with Max Delbrück and Milislav Demerec. 1945. "Types of Morphology Found in Bacterial Viruses." *Journal of Applied Physics* 16 (1945): 264.

Andrade, E. N. da C. 1964. *Rutherford and the Nature of the Atom.* New York: Doubleday.

Avery, Oswald, C. M. MacLeod, and Maclyn McCarty. 1944. "Studies on the Chemical Nature of the Substance Inducing Transformation of Pneumococcal Types. Induction of Transformation by a Desoxyribonucleic Acid Isolated from Pneumococcus Type III." *Journal of Experimental Medicine* 37 (1944): 137–58.

Baade, Walter, with Fritz Zwicky. 1934. "On Supernovae." *Proceedings of the National Academy of Sciences* 20 (1934): 254–59.

Babel, Isaac. 1994. *Odessa Tales: The Collected Stories of Isaac Babel.* London: Penguin Press.

Beckett, Samuel. 1955. *Molloy.* New York: Grove Press.

Berg, Paul, with Maxine Singer. 2003. *George Beadle, An Uncommon Farmer: The Emergence of Genetics in the 20th Century.* Plainview, N.Y.: Cold Spring Harbor Laboratory Press.

Bernstein, Jeremy, and Gerald Feinberg, eds. 1986. *Cosmological Constants: Papers in Modern Cosmology.* New York: Columbia University Press.

Bethe, Hans. 1939. "Energy Production in Stars." *Physical Review* 55 (1939): 434–56.

———. 1966. Oral history transcript. American Institute of Physics. Interview by Charles Weiner and Jagdish Mehra at Cornell University on Oct. 27, 1966.

Bethge, Eberhard. 1970. *Dietrich Bonhoeffer: A Biography.* New York: Harper-Collins.

Blumberg, Stanley, with Gwinn Owens. 1976. *Energy and Conflict: The Life and Times of Edward Teller.* New York: G. P. Putnam's Sons.

Blumberg, Stanley, and Louis Panos. 1990. *Edward Teller.* New York: Charles Scribner's Sons.

Bohr, Niels. 1933. "Light and Life." *Nature* 131 (1933): 421–23, 457–59.

———. 1972. *Collected Works.* Amsterdam: North-Holland (11 volumes already published, including some correspondence and commentary).

Brenner, Sydney. 1974. "The Genetics of *Caenorhabditis elegans.*" *Genetics* 77 (1974): 71–94.

———. 2007. "Manners Good and Bad." *Science* 318: 1245–46.

Cairns, John, with Gunther Stent and James Watson, eds. 1966. *Phage and the Origins of Molecular Biology.* Plainview, N.Y.: Cold Spring Harbor Laboratory Press.

Calaprice, Alice. 2005. *The New Quotable Albert Einstein.* Princeton, N.J.: Princeton University Press.

Carlson, Elof. 1981. *Genes, Radiation, and Society: The Life and Work of H. J. Muller.* Ithaca, N.Y.: Cornell University Press.

Carroll, Lewis. 2009. *Through the Looking Glass.* New York: Classic Books International.

Casimir, Hendrik. 1983. *Haphazard Reality: Half a Century of Physics.* New York: Harper and Row.

Chandrasekhar, Subrahmanyan. 1977. Oral history transcript. American Institute of Physics. Interview by Spencer Weart at the University of Chicago on May 17, 1977.

Chandrasekhar, Subrahmanyan, with George Gamow and Merle Tuve. 1938. "The Problem of Stellar Energy." *Nature* 141 (1938): 27.

Christianson, Gale. 1995. *Edwin Hubble: Mariner of the Nebulae.* Chicago: University of Chicago Press.

Crick, Francis. 1988. *What Mad Pursuit: A Personal View of Scientific Discovery.* New York: Basic Books.

———.1994. *The Astonishing Hypothesis: The Scientific Search for the Soul.* New York: Charles Scribner's Sons.

Croswell, Ken. 1995. *The Alchemy of the Heavens: Searching for Meaning in the Milky Way.* New York: Anchor Books.

Delbrück, Max. 1928. "Ergänzung zur Gruppentheorie der Terme." *Zeitschrift für Physik* 51 (1928): 181–87.

———.1930. "Quantitatives zur Theorie der homöopolaren Bindung." *Annalen Physik* 5 (1930): 36–38.

———. 1940. "The Growth of Bacteriophage and Lysis of the Host." *Journal of General Physiology* 23 (1940): 643–60.

———. 1945. "What Is Life?" *Quarterly Review of Biology* 20 (1945): 370–72.

———.1948. "Biochemical Mutants of Bacterial Viruses." *Journal of Bacteriology* 56 (1948): 1–16.

———. 1949. "A Physicist Looks at Biology." *Transactions of the Connecticut Academy of Arts and Sciences* 38 (1949): 173–90.

———. 1970. "A Physicist's Renewed Look at Biology: Twenty Years Later." *Science* 168 (1970): 1312–15.

———. 1972. "Homo Scientificus According to Beckett." In W. Beranek, ed. *Science, Scientists and Society*. New York: Bogden and Quigley, pp. 132–52.

———. 1980. "Was Bose-Einstein Statistics Arrived at by Serendipity?" *Journal of Chemical Education* 57 (1980): 467–70.

———. 1986. *Mind from Matter? An Essay on Evolutionary Epistemology*. Palo Alto, Calif.: Blackwell Scientific Publishing.

Delbrück, Max, with Emory Ellis. 1939. "The Growth of Bacteriophage." *Journal of General Physiology* 22 (1939): 365–84.

Delbrück, Max, with George Gamow. 1931. "Übergangswahrscheinlichkeiten von angeregten Kernen." *Zeitschrift für Physik* 72 (1931): 492–99.

Delbrück, Max, with Linus Pauling. 1940. "The Nature of Intermolecular Forces Operative in Biological Processes." *Science* 92 (1940): 77–79.

Delbrück, Max, with Nikolai Timofeeff-Ressovsky and Kurt G. Zimmer. 1935. "Über die Natur der Genmutation und der Genstruktur." *Nachrichten von der Gessellschaft der Wissenschaften zv Göttingen: Matematisch Physische Klasse Fachgruppe* 6, no. 13 (1935): 190–245.

D'Herelle, Félix. 1926. *The Bacteriophage and Its Behavior.* Baltimore: Williams and Wilkins.

Dicke, Robert, James Peebles, Peter Roll, and David Wilkinson. 1965. "Cosmic Blackbody Radiation." *Astrophysical Journal* 142 (1965): 414–19.

Dinesen, Isak. 1934. *Seven Gothic Tales*. New York: Random House.

Dirac, Paul. 1929. "The Quantum Mechanics of Many-Electron Systems." *Proceedings of the Royal Society* (London) A123 (1929): 714–33.

———. 1937. "The Cosmological Constants." *Nature* 139 (1937): 323.

———. 1938. "A New Basis for Cosmology." *Proceedings of the Royal Society* (London) A165 (1938): 198–208.

———. 1958. *Quantum Mechanics*. 4th ed. Oxford: Oxford University Press.

Dyson, Freeman. 1967. "Time Variation of the Charge of the Proton." *Physical Review Letters* 19 (1967): 1291.

———. 1985. *Origins of Life*. Cambridge, UK: Cambridge University Press.

Einstein, Albert. 1922. *Naturwissenschaften* 10 (1922): 184. This is Einstein's
   review of Pauli's 1922 article on relativity theory, reprinted in Pauli (1964).
Enz, Charles. 2002. *No Time to Be Brief: A Scientific Biography of Wolfgang
   Pauli.* New York: Oxford University Press.
Eve, A. S. 1939. *Rutherford.* Cambridge, UK: Cambridge University Press.
Fermi, Enrico. 1962. *Collected Works.* Chicago: University of Chicago Press.
Fischer, Ernst Peter, and Carol Lipson. 1988. *Thinking About Science: Max
   Delbrück and the Origins of Molecular Biology.* New York: W. W. Norton.
French, Anthony, and P. J. Kennedy. 1985. *Niels Bohr: A Centenary Volume.*
   Cambridge, Mass.: MIT Press.
Frisch, Otto. 1979. *What Little I Remember.* Cambridge, UK: Cambridge
   University Press.
Gamow, George. 1928. "The Quantum Theory of Nuclear Disintegration."
   Nature 122 (1928): 805–6.
———. 1931. *The Constitution of Atomic Nuclei and Radioactivity.* Oxford:
   Clarendon Press.
———. 1935. "Nuclear Transformations and the Origin of the Chemical
   Elements." *Ohio Journal of Science* 35 (1935): 1–7.
———. 1939. *Mr. Tompkins in Wonderland.* Cambridge, UK: Cambridge
   University Press.
———. 1940. *The Birth and Death of the Sun.* New York: Viking.
———. 1946. "Expanding Universe and the Origin of Elements." *Physical
   Review* 70 (1946): 572–73.
———. 1947. *One, Two, Three . . . Infinity.* New York: Viking.
———. 1948. "The Origin of the Elements and the Separation of
   Galaxies." *Physical Review* 74 (1948): 505–6.
———. 1952. *The Creation of the Universe.* New York: Viking.
———. 1952a. "The Role of Turbulence in the Evolution of Galaxies."
   Physical Review 86 (1952): 251.
———. 1954. "Possible Relation Between Deoxyribonucleic Acids and
   Protein Structures." *Nature* 173 (1954): 318.
———. 1964. *A Star Called the Sun.* New York: Viking.
———. 1966. *Thirty Years That Shook Physics: The Story of Quantum Theory.*
   New York: Doubleday. Reprint, New York: Dover: 1985.
———. 1967. "Electricity, Gravity, Cosmology." *Physical Review Letters* 19
   (1967): 759–61.
———. 1967a. "History of the Universe." *Science* 158 (1967): 766–69.
———. 1968. "Observational Properties of the Homogeneous and Isotropic
   Expanding Universe." *Physical Review Letters* 20 (1968): 1310–12.
———. 1970. *My World Line: An Informal Autobiography.* New York: Viking.
Gamow, George, with Mario Schönberg. 1941. "Neutrino Theory of Stellar
   Collapse." *Physical Review* 59 (1941): 539–47.

Garrity, Peter. 2007. *The Galloping Gamows: In Living Technicolor*. Documentary, available at www.booksurge.com.

Gingerich, Owen. 2004. "The Summer of 1953: A Watershed for Astrophysics." *Physics Today* (Dec. 2004): 34–40.

Gorelik, Gennady, and Victor Frenkel. 1994. *Matvei Petrovich Bronstein and Soviet Theoretical Physics in the Thirties*. Basel, Switzerland: Birkhauser.

———. 1995. "The Top Secret Life of Lev Landau." *Scientific American* (Aug. 1995): 72–77.

———. 1997. "Lev Landau, Pro-socialist Prisoner of the Soviet State." *Scientific American* (May 1997): 11–15.

Greenspan, Nancy Thorndike. 2005. *The End of the Certain World: The Life and Science of Max Born*. New York: Basic Books.

Greenspan, Neil. 2001. "You Can't Have It All." *Nature* 409 (2001): 137.

Gribbin, John. 1986. *In Search of the Big Bang: Quantum Physics and Cosmology*. New York: Bantam Books.

Guth, Alan. 1997. *The Inflationary Universe: The Quest for a New Theory of Cosmic Origins*. Reading, Mass.: Helix Books.

Harper, Eamon. 2001. "George Gamow: Scientific Amateur and Polymath." *Physics in Perspective* 3 (2001): 203–27.

Harper, Eamon, with W. C. Parke and G. D. Anderson, eds. 1997. *The George Gamow Symposium*. San Francisco: Astronomical Society of the Pacific.

Hawkes, Peter, ed. 2010. *Advances in Imaging and Electron Physics*. New York: Academic Press.

Hayes, William. 1993. "Max Delbrück." In *Biographical Memoirs of the National Academy of Sciences*. Washington, D.C.: National Academies Press, vol. 60 (1993): 67–117.

Heilbron, John. 1986. *The Dilemmas of an Upright Man: Max Planck as Spokesman for German Science*. Berkeley: University of California Press.

Hentschel, Klaus, ed. 1996. *Physics and National Socialism*. Basel, Switzerland: Birkhauser.

Hershey, Alfred. 1981. Max Delbrück Laboratory Dedication Ceremony. Cold Spring Harbor, N.Y.

Hoyle, Fred. 1950. *The Nature of the Universe*. Oxford: Basil Blackwell.

———. 1994. *Home Is Where the Wind Blows: Chapters from a Cosmologist's Life*. Oxford: Oxford University Press.

Hufbauer, Karl. 2007. "Landau's Youthful Sallies into Stellar Theory: Their Origins, Claims and Receptions." *Historical Studies in the Physical and Biological Sciences* 37 (2007): 337–54.

———. 2009. "George Gamow." In *Biographical Memoirs of the National Academy of Sciences*. Washington, D.C.: National Academies Press, pp. 3–39.

Inglis, John, with Joseph Sambrook and Jan Witkowski. 2003. *Inspiring Science: Jim Watson and the Age of DNA*. Plainview, N.Y.: Cold Spring Harbor Laboratory Press.

Jacob, François. 1982. *The Logic of Life*. New York: Random House.

———. 1998. *Of Flies, Mice and Men*. Cambridge, Mass.: Harvard University Press.

Johnson, George. 2005. *Miss Leavitt's Stars*. New York: W. W. Norton.

Judson, Horace Freeland. 1996. *The Eighth Day of Creation: Makers of the Revolution in Biology*. Exp. ed. Plainview, N.Y.: Cold Spring Harbor Laboratory Press.

Jungck, Robert. 1956. *Brighter Than a Thousand Suns: A Personal History of the Atomic Scientists*. New York: Harcourt.

Kahn, Charles. 2001. *Pythagoras and the Pythagoreans: A Brief History*. Indianapolis: Hackett Publishing Company.

Kaiser, David. 2007. "The Other Evolution Wars." *American Scientist* 95 (2007): 518–25.

Kay, Lily. 1993. *The Molecular Vision of Life: Caltech, the Rockefeller Foundation, and the Rise of the New Biology*. New York: Oxford University Press.

———. 2000. *Who Wrote the Book of Life? A History of the Genetic Code*. Stanford, Calif.: Stanford University Press.

Keller, Evelyn Fox. 2000. *The Century of the Gene*. Cambridge, Mass.: Harvard University Press.

Kelly, Cynthia. 2007. *The Manhattan Project*. New York: Black Dog and Leventhal.

Khalatkinov, Isaak M., ed. 1989. *Landau: The Physicist and the Man*. Oxford: Pergamon Press.

Koestler, Arthur. 1989. *The Sleepwalkers*. London: Arkana.

Knowlson, James. 1996. *Damned to Fame: The Life of Samuel Beckett*. New York: Grove Press.

Kragh, Helge. 1979. "Niels Bohr's Second Atomic Theory." *Historical Studies in the Physical Sciences* 10 (1979): 123–86.

———. 1996. *Cosmology and Controversy: The Historical Development of Two Theories of the Universe*. Princeton, N.J.: Princeton University Press.

———. 1999. *Quantum Generations: A History in the 20th Century*. Princeton, N.J.: Princeton University Press.

Lightman, Alan, and Roberta Brawer. 1990. *Origins: The Lives and Worlds of Modern Cosmologists*. Cambridge, Mass.: Harvard University Press.

Livanova, Anna. 1980. *Landau: A Great Physicist and Teacher*. New York: Oxford University Press.

Luria, Salvador E. 1984. *A Slot Machine, A Broken Test Tube: An Autobiography*. New York: Harper and Row.

Luria, Salvador E., and Max Delbrück. 1943. "Mutations of Bacteria from Virus Sensitivity to Virus Resistance." *Genetics* 28 (1943): 491–511.

McCarty, Maclyn. 1985. *The Transforming Principle: Discovering That Genes Are Made of DNA*. New York: W. W. Norton.

McKaughan, Daniel. 2005. "The Influence of Niels Bohr on Max Delbrück." *Isis* 96 (2005): 507–29.

McLaughlin, P. J. 1957. *The Church and Modern Science*. Available at Philosophical Library, www.philosophicallibrary.com.

Mather, John, with John Boslough. 2008. *The Very First Light: The True Inside Story of the Scientific Journey Back to the Dawn of the Universe*. New York: Basic Books.

Medawar, Peter. 1968. "Lucky Jim." *New York Review of Books* (Mar. 28, 1968): 3–5.

Medvedev, Zhores. 1982. "Nikolai Wladimovich Timofeef-Ressovsky (1900–1981)." *Genetics* 100 (1982): 1–16.

Misner, Charles, with Kip Thorne and John Wheeler. 1973. *Gravitation*. San Francisco: W. H. Freeman.

Moore, Walter. 1989. *Schrödinger, Life and Thought*. Cambridge, UK: Cambridge University Press.

Morange, Michel. 1998. *A History of Molecular Biology*. Cambridge, Mass.: Harvard University Press.

Olby, Robert. 1974. *The Path to the Double Helix: The Discovery of DNA*. New York: Dover.

Overbye, Dennis. 1991. *Lonely Hearts of the Cosmos: The Scientific Quest for the Heart of the Universe*. New York: HarperCollins.

Pais, Abraham. 1982. *Subtle Is the Lord: The Life and Science of Albert Einstein*. New York: Oxford University Press.

———. 1991. *Niels Bohr's Times, in Physics, Philosophy, and Polity*. New York: Oxford University Press.

Pauli, Wolfgang. 1964. *Scientific Correspondence*. 2 vols. K. von Meyenn, R. Kronig, and V. Weisskopf, eds. New York: Interscience.

———. 1994. *Scientific Correspondence*. Vol. 3. K. von Meyenn, ed. Berlin: Springer-Verlag.

Penzias, Arno, and Robert Wilson. 1965. "A Measurement of Excess Antenna Temperature at 4080 Mc/s." *Astrophysical Journal* 142 (1965): 419–20.

Perlmutter, Saul. 2003. "Supernovae, Dark Energy and the Accelerating Universe." *Physics Today* (Apr. 2003): 53–60.

Perutz, Max. 1989. *Is Science Necessary? Essays on Science and Scientists*. New York: E. P. Dutton.

———. 1998. *I Wish I'd Made You Angry Earlier: Essays on Science and Scientists*. Plainview, N.Y.: Cold Spring Harbor Laboratory Press.

Pringle, Peter. 2008. *The Murder of Nikolai Vavilov: The Story of Stalin's Persecution of One of the Great Scientists of the Twentieth Century*. New York: Simon and Schuster.

Raffelt, Georg. 1996. *Stars as Laboratories for Fundamental Physics.* Chicago: University of Chicago Press.

Ratner, Vadim. 2001. "Nikolay Vladimirovich Timofeeff-Ressovsky (1900–1981): Twin of the Century of Genetics." *Genetics* (2001) 158: 933–39.

Reines, Frederick, ed. 1972. *Cosmology, Fusion and Other Matters: George Gamow Memorial Volume.* Boulder, Colo.: Colorado Associated Press.

Rhodes, Richard. 1988. *The Making of the Atom Bomb.* New York: Simon and Schuster.

Rich, Alex. 1995. "The Nucleic Acids: A Backward Glance." *Annals of the New York Academy of Sciences* 758 (1995): 97–142.

———. 2009. "The Era of RNA Awakening: Structural Biology of RNA in the Early Years." *Quarterly Review of Biophysics* 42 (2009): 117–37.

Rubin, Vera. 2002. "Intuition and Inspiration Made Gamow a Star Turn." *Nature* 415 (2002): 13.

Salpeter, Edwin. 1996. "Reminiscences of George Gamow and Nuclear Astrophysics." *Astronomical and Astrophysical Transactions* (1996): 3–37.

Schrödinger, Erwin. 1944. *What Is Life? The Physical Aspect of the Living Cell.* Cambridge, UK: Cambridge University Press.

Schweber, Silvan. 2000. *In the Shadow of the Bomb.* Princeton, N.J.: Princeton University Press.

Segrè, Gino. 2000. "The Big Bang and the Genetic Code." *Nature* 404 (2000): 437.

———. 2007. *Faust in Copenhagen: A Struggle for the Soul of Physics.* New York: Viking.

Shakespeare, William. 1992. *The Merchant of Venice.* Folger Shakespeare Library. New York: Simon and Schuster.

Shropshire, Walter, Jr., ed. 2007. *Max Delbrück and the New Perception of Biology.* Bloomington, Ind.: AuthorHouse.

Sime, Ruth. 1996. *Lise Meitner: A Life in Physics.* Berkeley: University of California Press.

Singh, Simon. 2004. *Big Bang: The Most Important Scientific Discovery of All Time and Why You Need to Know About It.* London: Fourth Estate.

Sloan, Phillip R., and Brandon Fogel. Forthcoming 2011. *Creating a Physical Biology: The Three-Man Paper and Early Molecular Biology.* Chicago: University of Chicago Press.

Smoot, George, and Keay Davidson. 1994. *Wrinkles in Time: Witness to the Birth of the Universe.* New York: William Morrow.

Spruch, Grace Marmor. 1979. "Pyotr Kapitza: Octogenarian Dissident." *Physics Today* (Sept. 1979): 34–41.

Stahl, Frank. 2001. "Alfred Hershey." In *Biographical Memoirs of the National Academy of Sciences.* Washington, D.C.: National Academies Press, vol. 80, pp. 142–59.

Stent, Gunther. 1968. "That Was the Molecular Biology That Was." *Science* 160 (1968): 390–95.

———. 1981. "Obituary: Max Delbrück." *Trends in Biochemical Sciences* 6 (1981): iii–iv.

Stuewer, Roger H. 1994. "The Origin of the Liquid-Drop Model and the Interpretation of Nuclear Fission." *Perspectives on Science* 2 (1994): 76–129.

Summers, William C. 1999. *Félix d'Herelle and the Origins of Molecular Biology.* New Haven, Conn.: Yale University Press.

Thorne, Kip. 1994. *Black Holes and Time Warps: Einstein's Outrageous Legacy.* New York: W. W. Norton.

Tropp, Eduard. 1993. *Alexander Friedmann: The Man Who Made the Universe Expand.* Cambridge, UK: Cambridge University Press.

Turner, Michael S. 2008. "From Alpha, Beta, Gamma to Precision Cosmology: The Amazing Legacy of a Wrong Paper." *Physics Today* (Dec. 2008): 8.

Tuve, Merle. 1982. Oral history transcript. American Institute of Physics. Interview by Thomas Cornell at Chevy Chase, Md., on Jan. 13, 1982.

Vallentin, Antonina. 1954. *Einstein: A Biography.* London: Weidenfeld and Nicolson.

Van der Waerden, Bartel. 1967. *Sources of Quantum Mechanics.* Amsterdam: North-Holland.

Von Meding, Dorothee. 1997. *Courageous Hearts: Women and the Anti-Hitler Plot of 1944.* Providence, R.I.: Berghahn Books.

Von Neumann, John. 1961. *Collected Works.* Vol. 5. A. H. Taub, ed. New York: Macmillan, pp. 288–328.

Wali, Kameshwar. 1984. *Chandra: A Biography of S. Chandrasekhar.* Chicago: University of Chicago Press.

Watson, James D. 1968. *The Double Helix: A Personal Account of the Discovery of the Structure of DNA.* New York: Athenaeum.

———. 2000. *A Passion for DNA: Genes, Genomes, and Society.* Plainview, N.Y.: Cold Spring Harbor Laboratory Press.

———. 2001. *Genes, Girls, and Gamow: After the Double Helix.* New York: Alfred A. Knopf.

Watson, James D., with Andrew Berry. 2003. *DNA: The Secret of Life.* New York: Alfred A. Knopf.

Watson, James D., and F. H. C. Crick. 1953. "A Structure for Deoxyribose Nucleic Acid." *Nature* 171 (Apr. 25): 737–38.

———. 1953a. "Genetical Implications of the Structure of Deoxyribonucleic Acid." *Nature* 171 (May 30): 964–67.

Weart, Spencer, and Gertrude Weiss Szilard, eds. 1978. *Leo Szilard: His Version of the Facts—Selected Recollections and Correspondence.* Cambridge, Mass.: MIT Press.

Weinberg, Robert. 1993. *The Revolution of 1905 in Odessa: Blood on the Steps.* Bloomington: Indiana University Press.

Weinberg, Steven. 1977. *The First Three Minutes: A Modern View of the Origin of the Universe.* New York: Basic Books.

———. 2001. *Facing Up: Science and Its Cultural Adversaries.* Cambridge, Mass.: Harvard University Press.

———. 2005. "Einstein's Mistakes." *Physics Today* (Nov. 2005): 31–35.

———. 2009. *Lake Views: This World and the Universe.* Cambridge, Mass.: Harvard University Press.

Wenkel, Simone, and Ute Deichmann, eds. 2007. *Max Delbrück and Cologne: An Early Chapter of German Molecular Biology.* Singapore: World Scientific.

Wiener, Jonathan. 1999. *Time, Love, Memory: A Great Biologist and His Quest for the Origins of Human Behavior.* New York: Vintage Books.

Zee, Anthony. 2003. *Quantum Field Theory in a Nutshell.* Princeton, N.J.: Princeton University Press.

Zen Buddhism: An Introduction to Zen with Stories and Parables. 1959. Mount Vernon and New York: Peter Pauper Press.

Index

Page numbers *in italics* refer to illustrations.

Bragg, Sir Lawrence, 199–200
Brahe, Tycho, 276
brain, 271–74
Brecht, Bertolt, 89
Brenner, Sydney, 141–42
Broad Institute, 269
Bronstein, Matvei, 294n
Bruce, Mary Adeline, see Delbrück, Mary
    Adeline Bruce
Brunelleschi, Filippo, 21
Burke, Bernard, 241–42

California, 121, 153–54, 191, 257,
    270–71
California Institute of Technology
    (Caltech), xix, 122, 189, 203–4,
    207, 221, 225, 261, 274, 287
  Beadle at, 165, 172, 173
  Benzer at, 229
  Biology Division of, 165, 173
  Delbrück at, 93–94, 96–101, 117, 144,
      162–63, 165, 170, 172, 173–74, 177,
      192, 230, 232–33, 261, 271–72
  Delbrück's views on, 172, 173
  Gamow's visit to, 215
  Pauling at, 165, 173, 174, 215
Cambridge University, 19, 28–30,
    36–37, 40, 44, 68, 69, 174, 185–86,
    199–201, 203, 205, 221–23, 304n
  Chandrasekhar at, 80, 81
Cambridge University Press, 279, 283
camping, 171, 175–76, 276
Canada, 185, 200
cancer, 99, 101, 285
Cape Cod, 216–17, 304n
Capra, Frank, 154
carbon, 84, 86, 123, 188–89, 190
Carlsberg Fellowship, 28, 29
Carnegie Institution, 107, 191, 241
Carroll, Lewis, 278
Casimir, Hendrik, 38–39
Casino da Urca, 111, 115
Catherine the Great, 12
Catholic Church, 155, 187
Cavendish Laboratory, 199–200
Cepheid variables, 153, 154, 257
Cerdá-Olmedo, Enrique, 228, 232
Chandrasekhar, Subrahmanyan
    "Chandra," 80–83, 114, 301n
Chandrasekhar limit, 80
Chargaff, Erwin, 197
Chargaff's rules, 197, 210
Chase, Martha, 198
chemistry, 56, 99, 104, 105, 135, 159,
    163, 164, 173

quantum mechanics applied to,
    32–33, 205
  see also biochemistry
Chicago, University of, 82, 134, 160
chromosomes, 58, 128, 193
Churchill, Sir Winston, 178
"clear paradoxes," 231
Clinton, Bill, 267
COBE (Cosmic Background Explorer),
    249–51, 249
Cockcroft, John, 24, 30
cold load, 240
Cold Spring Harbor Laboratory, xix, 96,
    121, 122, 137–43, 170, 171, 173,
    177, 192, 203, 225, 230
  Delbrück's Festschrift at, 228, 265,
      266, 270, 272–73
  phage course at, 137–42, 160, 228, 229
  symposium at (1946), 142–43
  symposium at (1953), 206, 207
Cologne, 225–27
Colorado, University of, 224, 278, 287
Columbia University, 58, 93, 96, 103,
    106, 107, 120, 161, 197
Coma galaxy, 253
Communist Party, Soviet, 63, 64
Communists, 88, 295n
complementarity, 46, 50–51, 54,
    62, 134
computer science, 163, 164, 273
Connecticut Academy of Arts and
    Sciences, 230–31
consciousness, 271–74
Copenhagen, xvii, 18, 21, 42, 48, 68,
    101–2, 159, 171, 176, 227, 262
  boardinghouse in, 38, 39, 80–81
  Bohr's Institute in, see Institut for
      Teoretisk Fysik
  Bohr's Rigsdag lecture in, 52–54
  Caltech compared with, 177, 230
  Cold Spring Harbor compared with,
      139, 230
Copenhagen, University of, 22
correspondence principle, 21
cosmic microwave background radiation
    (CMBR), 244–51, 306n
cosmological constant, 150, 155, 253–61
cosmology, xv–xvii, 146–60, 216, 219,
    220, 238–65, 286
  Big Bang, see Big Bang cosmology
  CMBR and, 244–51, 306n
  Creation of the Universe, The (Gamow),
      187–88, 280
Crick, Francis, xvii, xix, 135, 213–17,
    219–22, 264, 286, 304n

Rous, Peyton, 99
Royal Astronomical Society, 155
Royal Danish Academy of Sciences, 28, 214
Royal Society, British, 27, 69
Rubin, Vera, 185, 191, 253
Rundle, Robert, 222–23
Russia, 2, 12–28, 43, 63–74, 144
    Gamow's love for, 64, 144, 177–78
    revolt (1905) in, 13–14
    tsarist, 12–14
    see also Soviet Union
Russian Army, xx, 13, 16, 110
Russian Orthodox Church, 13, 14–15
Russian Revolution (1917), 14
Russo-Japanese War (1905), 13
Rutherford, Ernest, 19–20, 22, 28–30, 59, 68, 199
    atomic model of, 9, 165, 174, 204
    Kapitza's relationship with, 69, 70

St. Petersburg (Petrograd; Leningrad), 12, 15–18, 28, 65
Salk Institute, 270–71
Salpeter, Edwin, 188–90
Sanger, Fred, 266–67
Schoenberg, Arnold, 89
Schönberg, Mario, 112, 115–16
Schrödinger, Erwin, 9, 17, 24–25, 53, 61, 131–35, 164, 261, 272
Science, 252, 262, 302n
Segrè, Bettina, 307n
Segrè, Julie, 268–69
sensory physiology, 231–33
Slipher, Vesto, 153, 154
Smoot, George, 250–51
Snow, Charles Percy, 279
Solvay Conference (1933), 67–68
Solzhenitsyn, Alexander, 297n
Sommerfeld, Arnold, 3, 170
Soviet Union:
    Gamow's escape from, 64–70, 72, 86, 92, 295n
    Gamow's return to, 41–42, 63
    genetics in, 59, 72
    German zone of, 166, 167
    purges in, 294n, 297n
    secret police in, 63, 64, 71, 295n
    under Stalin, xx, 42, 63, 69–72, 87, 295n
    see also Russia
space, 148, 154
Spencer Jones, Sir Harold, 187
sperm cell nuclei, 198
Stahl, Frank, 207, 304n

Stalin, Joseph, xx, 42, 63, 69–72, 87
    Hitler's pact with, 295n
    Kapitza's correspondence with, 70–71
    purges of, 69, 294n, 297n
Star Called the Sun, A (Gamow), 280
stars, 8–9, 80–85, 153–54, 156–57, 253
    energy source for cores of, 30–31, 77–78, 81–84
    neutron, 114, 115, 253
    novae, 8, 113, 114, 115, 117
    signals emitted by, 239–40
    supernovae, 111–15, 188, 257, 299n
Stars as Laboratories for Fundamental Physics (Raffelt), 112
statistical mechanics, 55, 247, 284
Stent, Gunther, 124, 215, 270, 272–73
Stockholm, 234–36
Strassmann, Fritz, 104–7, 109
Strauss, Lewis, 108
Strömgren, Bengt, 77, 82
"Structure and Properties of Atomic Nuclei," 67–68
Sturtevant, Alfred, 58, 97
sun, 77, 81–85, 111, 113, 149, 157, 255, 276, 280
Supernova Cosmology Project, 257
supernovae, 111–15, 188, 257, 299n
superstring theory, 260
Sweden, 102, 104–6, 234–36
Switzerland, 39, 44–47, 118
Szent-Györgi, Albert, 304n
Szilard, Leo, 10, 107–11, 117, 137–38, 160–61, 206

Tatum, Edward, 122–23, 134, 143, 172
    "one gene, one enzyme" theory of, 123, 142, 162, 163, 197, 264
telescopes, 14, 153, 239–40, 244
Teller, Edward, 74–78, 82, 92, 103, 109, 110, 130, 156–57, 216, 262
    on Gamow, 84, 201
Teller, Mici, 75–76
textbooks, 36–37, 112, 279
thermal equilibrium, 244, 247–48
thermal radiation, 238–42
thermodynamics, 247
Thirty Years That Shook Physics (Gamow), 43, 280
Three Musketeers, 17–18, 294n
Through the Looking Glass (Carroll), 278
Timofeev-Ressovsky, Nikolai, 55, 57–63, 69, 91, 93
    Delbrück's collaboration with, 60–62, 85, 92, 97, 119, 132
    after World War II, 72–73, 297n